An Urchin in the Storm

BY STEPHEN JAY GOULD IN
NORTON PAPERBACK

EVER SINCE DARWIN
Reflections in Natural History

THE PANDA'S THUMB
More Reflections in Natural History

THE MISMEASURE OF MAN

HEN'S TEETH AND HORSE'S TOES
Further Reflections in Natural History

THE FLAMINGO'S SMILE
Reflections in Natural History

AN URCHIN IN THE STORM
Essays about Books and Ideas

ILLUMINATIONS: A BESTIARY (with R. W. Purcell)

WONDERFUL LIFE
The Burgess Shale and the Nature of History

BULLY FOR BRONTOSAURUS
Reflections in Natural History

An Urchin in the Storm

Essays about Books and Ideas

Stephen Jay Gould

W·W·NORTON & COMPANY
NEW YORK LONDON

Printed in the United States of America.

The text of this book is composed in Baskerville, with display type set in Baskerville.
Composition by the Haddon Craftsmen, Inc.

First published as a Norton paperback 1988

Library of Congress Cataloging-in-Publication Data
Gould, Stephen Jay.
An urchin in the storm: essays about books and ideas
by Stephen Jay Gould.
p. cm.
Includes index.
1. Biology. 2. Biology—Book reviews. I. Title.
QH311.G68 1987 87-21718
574—dc19 CIP

ISBN 0-393-30537-6

W.W. Norton & Company, Inc.
500 Fifth Avenue, New York, N.Y. 10110
W.W. Norton & Company, Ltd
10 Coptic Street, London WC1A 1PU

8 9 0

From the heartland of vestigial American anglophilia
To my two favorite British intellectuals—Arab and Jew by origin.
E pluribus unum.

For Peter Medawar, a man of consummate bravery, penetrating intellect, unsurpassed *joie de vivre,* and unfailing encouragement to men of good will; and
For Isaiah Berlin, the polymath of our time, who once befriended a young scholar with no status at all, and who gave old Archilochus his biggest boost since Erasmus.
In appreciation for their inspiration but, above all, for their kindness.

Contents

Preface 9

I *Evolutionary Theory*

1. How Does a Panda Fit? 19
2. Cardboard Darwinism 26
3. Misserving Memory 51
4. The Ghost of Protagoras 62

II *Time and Geology*

5. The Power of Narrative 75
6. Deep Time and Ceaseless Motion 93

III *Biological Determinism*

7. Genes on the Brain 107
8. Jensen's Last Stand 124
9. Nurturing Nature 145

IV *Four Biologists*

10. Triumph of a Naturalist 157
11. Thwarted Genius 169
12. Exultation and Explanation 180
13. Calling Dr. Thomas 189

V *In Praise of Reason*

14. Pleasant Dreams 199
15. The Perils of Hope 208
16. Utopia, Limited 216
17. Integrity and Mr. Rifkin 229
18. The Quack Detector 240

Index 247

Preface

The U.S. Army Teaching Manual is a compendium of wise and practical advice from an institution that surpasses all others (probably even the public schools, since most kids retain some spark of interest) in its need to foist information upon an unwilling and unresponsive audience. This manual lists, as its first rule, "never apologize." Yet I do not know how else to introduce a volume of book reviews (though my apology is only a tactical feint before the right cross of justification).

I once heard a speech by Herblock, addressed to aspiring journalists and dedicated to eradicating the notion that newspaper articles could be more than strictly ephemeral and eminently forgettable. He cited an old motto from the era before extra-duty plastic bags: yesterday's paper wraps today's garbage.

On the continuum from Rupert Murdoch to King James, old book reviews must, as a genre, fall at or very near the ephemeral end. A second argument against the preservation of book reviews lies in their central character—for no literary activity is so despised and excoriated (often properly so) as the reviewer's art. I picture several reviewers of my own books as passing a long future lodged between Brutus and Judas in the jaws of Satan. The famous quip of Max Reger to a music critic well illustrates the dubious propriety of reviewing as a general activity, the attitude of creative people to the genre, and the ephemerality (and place of final disposal) for its results: "I am sitting in the smallest room in the house. I have your review in front of me. Soon it will be behind me." Yet the very vehemence of dismissal also indicates—on the old Shakespearian principle of protesting too much—that reviews are not so lightly ignored. Charles Lamb proclaimed, "for critics I care the five hundred thousandth part of the tithe of a half-farthing." But why waste such emotional verbiage on an activity really accorded (if I calculate correctly) but one forty millionth of an old English penny in value.

Yet books are the wellspring and focus of our lives as scholars. Commentary upon such a source should, at its best, be expansive and enlightening—a sign of respect for a basic product. That so many book reviews are petty, pedantic, parochial, pedestrian (add your own p's and q's, querulous, quotidian, quixotic)—so much so that they have folded what might be an honorable genre into their gripping nastiness—strikes me as a sadness that might not lie beyond hope of reversal. Why should articles of commentary on other books not lie within the domain of the essay?

To gain such a status as potentially worth preserving, such commentaries would have to forego what many take to be the primary charge of a book review—to present a detailed account of the work's content and merit. (Most books, after all, are ephemeral; their specifics, several years later, inspire about as much interest as daily battle reports from the Hundred Year's War). Another type of book review—one that uses another writer's work as an anchor for discussing an issue of wider scope—might provoke an author's ire, but might also gain, by its generality, entry into the class of essays.

Since I am most moved by general themes, but find them vacuous unless rooted in some interesting particular, I have always tried to write book reviews in this broader style. I even dare to hope that some authors might be pleased to see their particular work used as a focus for general discussion, rather than "reviewed" by the traditional listing of likes and dislikes. (I am at least consistent in foisting upon others what I hope for myself. Among all reviews of my own books, I particularly cherish a long article, also from the *New York Review of Books,* on "Ontogeny and Phylogeny" written by the great British zoologist, J.Z. Young. He wrote a wonderfully perceptive essay on the relationship between embryology and evolution, but only bothered to mention that I had written a book about the subject in the last two paragraphs. Made me think (I am not being facetious) that I had done something worthwhile in choosing a meaty subject, long neglected. I would much rather be an interesting particular for an enduring generality than an item of, by and for itself and the moment alone. *Verweile doch, du bist so schön.*)

These essays are not quite so unmindful of their generating products as Young on ontogeny. We must believe that books, and their ideas, have an enduring meaning worth consideration after an actual title goes out of print (though nearly all discussed herein remain on the shelves of good bookstores, at least as paperbacks). Each of these chapters uses an individual book to pursue a general theme, but organizes its discussion as a critique of content. (I have also written some reviews in the more conventional mode of local judgment—but these I have not included.) All but one of these essays originally appeared in *The New York Review of Books.* My deepest thanks to Robert Silvers, a great editor who makes fine (and even insistent) suggestions but who will not change a comma without consulting. I hope that these pieces follow his goal for a genre of general commentary rooted in the review of books.

My second rationale for this collection lies in the coherence that I (at least) detect among its disparate subjects. I didn't impose any theme as I worked piecemeal over a decade, but a judicious choice of titles by Bob Silvers combined with a personal, stubborn consistency of viewpoint—the hobgoblin of small minds to be sure—combined to record a particular view of nature and human life: the perspective of an evolutionist committed to understanding the curious pathways of history as irreducible, but rationally accessible.

I don't think that coherence is an unmixed blessing, or necessarily a virtue at all in our complex world. It forces one to take the hedgehog's part in that overused (and basically uninterpretable) aphorism, attributed to Archilochus and kept alive through the ages by scholars from Erasmus to Isaiah Berlin: "The fox knows many things, but the hedgehog knows one big thing." Vulpine flexibility may be the greater virtue in such a diverse and dangerous world. Think what the real hedgehogs of history, Cortez and Pizarro for example, might have done with modern technologies of destruction.

Still, coherence—the way of the hedgehog—may be salutary for the harmless endeavor of forging a book from elements written separately and without thought of later conjunction. Also, the hedgehoggery of a natural historian must be the most benign form of the affliction that one could devise.

We would have more comfort in our strange and difficult world if we could believe (as our cultural traditions have tried so mightily to impose) that human mentality is the sensible and predictable result of a process directed towards this goal from its inception. (Since human history only graces the last second of the cosmic year, we feel even more driven to identify our inexorable antecedents in earlier history—lest we be forced to confront our status as a fortunate afterthought). Yet history, with its quirky pathways and quixotic reorganizations, teaches a hard lesson. Unless God is even more inscrutable than we ever dared to imagine (or unless He explicitly designed our modes of thought so that we would never grasp His own), the history of life confers no special or preordained status upon human intelligence.

Against this natural assault upon traditional hopes, the hedgehog takes his traditional posture: he rolls up into a ball, nose against anus, and thrusts his prickles against the world. Yet this metaphor, imposed by Archilochus himself, is too harsh and negative. The hedgehog's posture can be joyous and expansive. Like the porcupine of legend, he can aim his darts and strike like Cupid. (Real porcupines do no such thing, and are not closely related to hedgehogs in any case. The world, as I said, is never simple; it doesn't even provide apt metaphors.) The happy hedgehog of this volume roots his posture—and the organization of his essays—into three linked statements about nature and knowledge.

First, nature's way: If life's history cannot be read as an ascending ladder to human wisdom, step after predictable step, neither can the opposite pole of true randomness capture its evident order. Life's history is massively contingent—crucially dependent upon odd particulars of history, quite unpredictable and unrepeatable themselves, that divert futures into new channels, shallow and adjacent to old pathways at first, but deepening and diverging with the passage of time. We can explain the actual pathways after they unroll, but we could not have predicted their course. And if we could play the game of life again, history would roll down another set of utterly different but equally explainable channels. In this crucial sense, life's history does not work like the stereotype of a high-school physics experiment. Irreducible history is folded into the products of time.

Art has often understood this fundamental theme better than science. Marty McFly, trying to preserve the possibility of his own birth, struggles to reunite his parents in *Back to the Future.* In *It's a Wonderful Life* (now in color by coloration, as Frank Capra spins in his grave), the angel grants Jimmy Stewart his request to see how his town would have developed had he never been born. Stewart is astounded by the differences (all negative), and the angel replies that "each man's life touches so many other lives."

The first two sections of this book group reviews that discuss the irreducibility of history (and the pleasures and challenges of contingency) in its two principal domains of life and the earth. The first section on evolution focuses upon structuralist and historicist alternatives to what I regard as the mistaken functionalist paradigm of adaptation that still shapes Darwinian theory, while (in striking irony) ripping from that theory the very theme—history itself—that defines the subject of life. The second section on geology stresses the importance of narrative and historical uniqueness in setting a framework for the scientific study of time.

Second, explaining nature's complexity: We have traditionally matched our hopes for linear progress in nature with a mode of scientific explanation well suited to such simple systems: the reductionism of the Cartesian tradition, with its belief that complexity must be broken down (by "analysis") into constituent "atoms" that produce the phenomena of our scale along linear chains of causation regulated by laws of nature. I do not deny the power, or the great successes, of Cartesianism, but hold that its limits have probably been reached in the explanation of complex historical systems. This claim is no appeal to mysticism or intractability, but an argument that equally powerful (but different) techniques of historical *sciences*—with their themes of irreducible interaction, hierarchy, and resolvable contingency—must be embraced to break the hegemony of what, in our parochialism, we call *the* scientific method (the restricted domain of the Cartesian tradition, with its primary themes of experiment, laboratory control, repetition and quantification).

Section three of this book explores the theory and consequences (political and intellectual) of biological determinism, the primary Cartesian account of human nature (tabula rasa environmentalism would be equally reductionist—the issue is one of general explanatory approach, not specific claim). Section four discusses the life and work of four great biologists who struggled with life's contingency: two scientists (McClintock and Just) who based their work upon an intelligible holism rooted in the taxonomic approach to nature, and two representatives of the primary domains of contingency—natural history and medicine (Hutchinson and Thomas).

Third, a general plea for rationalism in explanation: A call against reductionism can attract strange and unwelcome bedfellows. Many people rightly sense the limits of Cartesianism but, still imbued with the cultural and psychological baggage of hope for a transcendent mind, both within our skull and in heaven above, decry rationality itself and make a false equation between the admitted wonder and uniqueness of human mentality and the necessity for mystical explanations that transcend the baseness of material reality. I maintain no hostility towards the hope for new principles in the explanation of mind, but rebel against the perilous slide from our current ignorance to a claim for ineffability (and the even further slide to a glorification of the nonrational). Complex, contingent, interactive, hierarchical do not mean unknowable—quite the reverse, for these are the tools of a different kind of rational understanding. No force can be so powerfully destructive, so capable of undoing the patient struggles of centuries with a single blow, than irrationalism (especially when fueled by the "true belief" that converts such fine concepts as patriotism and religion into dangerous weapons of destruction). Thus, the final section five, "in praise of reason," hovers around the precipice that pushes a rational stance towards complexity (the "New York holism" of my parochial response to Capra) over the edge into false hope, mysticism, and finally demagoguery. I discuss Dyson's pleasant but unsupportable reveries, Jastrow's misreading of evolution in the light of his theological hopes, Capra's flirtation with mysticism, Rifkin's rip-off, and, finally, Gardner's patient and happy sanity.

As a New Yorker who spent summers on Jones Beach and then studied invertebrate paleontology, I have a parochial attachment to creatures of the sea. I always wondered why globular, spiny echinoderms are called "sea urchins." I never could grasp their affinity with waifs of the city streets—until I discovered that hedgehogs are called urchins in Europe, and that the spiny exterior of these echinoderms does resemble a hedgehog rolled up against danger.

My title—pushing Archilochus even further—tries to capture several aspects of a naturalist's hedgehoggery (also my own background and personality). I stated that themes of antireductionism and historical contingency could define the most benign form of hedgehoggery. I say this because such a concept of nature awards the benefits of both the fox's and the hedgehog's world: the virtues of consistency in a view of life, with the wondrously diverse products—the vertitable storm of results—that an unpredictable contingency places upon our earth. We can revel in all the pretty pebbles for their own sake, while maintaining a coherent view of their estate.

The storm of my title also, and obviously, has a negative meaning—but do remember that the hedgehog's strategy of enrollment is no retreat or surrender: he presents a tough exterior, continues to prickle the enemy, and unrolls once again into glorious daylight. I do not underestimate the storm at several levels. Yet, as a card-carrying member of the guild

I

Evolutionary Theory

1

How Does a Panda Fit?

Many animals, including Jesse James, Alexander the Great, and the giant panda, must, Janus-like, show two faces to the world—one required by legend, the other given by nature. The hortatory faces are, in sequence, honest (in the largest sense), virtuous, and cuddly; the natural visages tend to thievery, rapacity, and ennui.

George B. Schaller and his colleagues, in the finest study yet completed on the second panda, write in their introduction:

> There are two giant pandas, the one that exists in our mind and the one that lives in its wilderness home. Soft, furry, and strangely patterned in black and white, with a large, round head and a clumsy, cuddly body, a panda seems like something to play with and hug. No other animal has so entranced the public. . . . The real panda, however, the panda as it lives in the wild, has remained essentially a mystery.

The Giant Pandas of Wolong, an attempt to decrease the mystery surrounding panda number two, provides extraordinary testimony to another phenomenon, more often part of legend than of fact—international cooperation in science. Only one thousand pandas or so survive in nature, all in six small blocks

A review of *The Giant Pandas of Wolong* by George B. Schaller, Hu Jinchu, Pan Wenshi, and Zhu Jing.

of bamboo forest (29,500 square kilometers) along the eastern edge of the Tibetan plateau—though historical records indicate a former distribution up to one thousand kilometers further east, nearly to the Pacific coast.

The Wolong Natural Reserve, largest of China's panda sanctuaries, holds between 130 and 150 animals. Chinese scientists began an in-depth study of Wolong pandas in 1978. George B. Schaller, from Wildlife Conservation International, arrived in December 1980 to work with a Chinese team headed by Hu Jinchu of Nanchong Normal College. *The Giant Pandas of Wolong* summarizes the joint work that continues today.

Since this book is about the second panda, it will rarely delight and charm. *The Giant Pandas of Wolong* is a technical treatise, not a contribution to the distinctive genre of popular books that describe a naturalist's intimate life with one interesting species in the wild (including several by Schaller, most notably his *Year of the Gorilla*). We can sense what's coming when we read on, page three (I shall provide a translation upon request) that "the zygomatic arches are spread widely, and the sagittal crest is prominent. . . . A typically carnivorous dentition (I^3_3 C^1_1 P^4_4 M^2_3 = 42, but P_1 may be absent) has been strongly modified for crushing and grinding food." And the relentless passive voice of conventional scientific prose imparts no charm or grace of composition, especially in such lines as "apparent itches are scratched with fore- or hindpaw."

Pandas are rare and elusive animals even in the relative abundance of their Wolong reserve. We dare not recognize them as the cute stuffed toys of our children; indeed, we must struggle mightily to see them at all. Between March 1978 and December 1980, Schaller and company saw pandas only sixteen times; the enlarged team recorded thirty-nine additional observations between January 1980 and May 1981. They write: "Most of our contacts were brief—a glimpse as an animal crossed an opening or ambled up a trail."

Researchers must therefore rely upon indirect methods, primarily two in this case—one old fashioned, the other newfangled. Fortunately, pandas defecate prodigiously, and with such regularity that number of droppings provides an adequate clock for time spent in any particular spot. It is, I suppose, a kind of ultimate dethronement for panda one (of legend) when we recognize that brówn cylinders, rather than furry bodies, form the major source of direct evidence for this study.

Schaller and his team then trapped six pandas and fitted them with radio collars. These sophisticated devices transmit different signals during times of activity and rest for pandas. The resulting data on geographic ranges and energy budgets indicate that pandas live in relatively small, well-defined areas, averaging just 4.5 square kilometers for females and 6.1 square kilometers for males, females tending to concentrate their activity within a smaller core area of the range, males roaming more widely.

During most (indeed nearly all) of their day, pandas just don't do anything calculated to inspire sustained human interest. Basically, they eat bamboo during active periods (about 60 percent of a day) and rest for the remaining 40 percent—all the while emitting the vast undigested bulk of their labor by the rear exit. Other activities—traveling, scent marking, and grooming, for example—consume only a percent or two of an average day. More, of course, happens during the mating season; Darwin's ultimate game of passing one's genetic heritage into future generations rarely passes without interest, energy, and (in most cases among animals of our ilk) strife.

In the midst of this bamboo-directed monotony, any peculiar burst of activity must kindle our excitement. Thus, we read with pleasure about the panda that stood on its hands and arched its back end up a tree for scent marking. And we almost shout for joy in learning that one subadult slid downhill (on chest and belly) when it could have walked in snow—and that, *mirabile dictu,* it once walked back uphill to do it again.

And yet, in a sense, I am glad that the life of pandas is so dull by human standards, for our efforts at conservation have little moral value if we preserve creatures only as human ornaments; I shall be impressed when we show solicitude for warty toads and slithering worms. If we continue to treasure the panda even when we learn that it will not return, in basic human delectation, the warmth and playfulness that we once inferred from its appearance, then we are well on our way to a proper respect for nature. (If we can then come to admire pandas for what they are, and even learn from them some of the lessons that nature's diversity always teaches, then we shall finally understand, and to our greatest benefit in both practical and spiritual terms, what Huxley called, in the language of his day, "man's place in nature.")

Moreover, the very monotony of panda behavior as bamboo-eating machines defines their major interest for evolutionary theory; Schaller's treatment of this central subject also provides my only major unhappiness with his fine book. Pandas, by evolutionary descent, are members of the order Carnivora—but they belie their name by subsisting almost entirely upon bamboo. Their ancestors once ate meat, but then switched to bamboo. By constraint of a heritage so contrary to their current life, pandas must struggle to process enough food. Their digestive apparatus is not well designed for herbivory. Schaller et al. specify three major reasons for the difficulty:

1. Pandas cannot digest bamboo leaves and stems efficiently. "The panda," they write, "has retained the simple digestive tract of a carnivore: it lacks a special chamber to retain food, and it has no symbiotic microbes to ferment cellulose into available nutrients."
2. Pandas must therefore derive nutrients from the easily digestible cellular contents and not from the valuable cell walls. (Pandas defecate so prodigiously because they cannot digest most of what they consume.)
3. Leaves and stems are mostly water and structural carbohydrates; pandas therefore obtain low nutritional return for amounts eaten.

Schaller's calculations show that pandas live on the very edge of sufficiency. They eat bamboo all day long because they must spend every waking hour at it in order to get enough for their low ratio of return to investment. An amusing insight into this marginality arises from Schaller's efforts to determine how many hours a panda must eat (at its observed rate of foraging, speed and size of bite, and value of food) in order to fuel its minimal requirements. His figure of 19.4 hours is impossibly high since pandas only averaged 15.4 hours of eating per day. (This calculation recalls another presented earlier in the book—that pandas defecate more than they eat.) Obviously, these calculations must leave something out (unless pandas subvert the laws of physics). Small increases in speed or bite size (or an occasional munching of two stems at once) would put pandas over their obvious edge of viability. But Schaller's effort does dramatically demonstrate that pandas, though surrounded by food, can barely extract enough from their bounty.

Nonetheless, Schaller's entire discussion proceeds within the prevailing adaptationist model. He interprets everything that pandas do as adaptations to their curious mode of life. He identifies, as the major goal of this study, an understanding of "how the panda is adapted to bamboo." In some trivial sense, of course, pandas are "adapted"—they are getting by. But this sense of adaptation has no meaning—for all animals must do well enough to hang in there, or else they are no longer with us. Simple existence as testimony to this empty use of adaptation is a tautology. Meaningful adaptation must be defined as actively evolved design for local circumstances, not mere muddling through with inherited features poorly suited to current needs.

Pandas do, of course, display a suite of *secondary* true adaptations to their primary, unsurmounted dilemma of trying to eat bamboo with a carnivore's digestive tract. They pick, prepare, and chew with efficiency that has actively evolved; they have even invented a famous false "thumb" to abet their struggle.[1] But surely, despite the conceptual cloak cast by adaptation over this book, the primary theme of panda life must be read as a shift of function poorly accommodated by a minimally altered digestive apparatus. When anatomical structures are co-opted for new functions from previous uses in a different past, we may not speak of adaptation. When, as with pandas,

[1]See my earlier book *The Panda's Thumb* (New York: W.W. Norton, 1980).

the co-opted organs work so precariously, appeals to adaptation are even less appropriate.

When, in unguarded moments, Schaller lets the conceptual blinders slip, he reports the panda's dilemma forcefully: "The longer food remains in the digestive tract, the more fully will it be utilized; thus a long intestine, as found in herbivores, might benefit the panda. . . . The panda's digestive tract lacks physical and physiological adaptations for processing a bulky, herbivorous diet." (Deer intestines may be fifteen times longer than their body, sheep twenty-five times; pandas rank with other carnivores in having intestines only four to seven times as long.) But allegiance to adaptation soon usurps any subtle discussion of history, and we return to Gleason's mode—"how sweet it is." The authors even argue that the wild panda's failure to accumulate body fat should be viewed as an adaptation to their stable food supply. (They mention that zoo pandas do store fat, so physiology does not preclude obesity.) Might I suggest the obvious alternative—that a little fat might be a good thing, but that pandas, eating all their waking day simply to get by, remain slim by constraint rather than design.

The debate about adaptation is not a petty, abstract nicety of academic life. It embodies our basic attitudes toward history. Evolutionary biology is the primary science of history; strict adaptationism, ironically, downgrades history to insignificance by viewing the organism's relation to environment as an isolated problem of current optimality. How inappropriate to clamp this conceptual lock upon the panda—a demonstration, if ever one existed, that past histories exert a quirky hold (through inefficiencies imposed by heritage) upon an imperfect present.

Writing so brilliantly about the hold of theory upon our ability to observe, Geoffroy Saint-Hilaire[2] stated in 1827: "At first useless, these facts had to remain unperceived until the

[2]Geoffroy was France's greatest structural morphologist of the nineteenth century. He fought many a battle with the adaptationists of his day. How appropriate that this general warning should come from a man also so committed to undoing the prejudices of adaptationism.

moment when the needs and progress of science provoked us to discover them." It is time to rescue history from the subverting power of Pangloss's spectacles.

Everyone knows that the panda's current plight extends well beyond the intrinsic dilemma of its unbreakable contract with bamboo (intensified by the tendency of most bamboo species to undergo mass flowering with subsequent death of edible plants and long periods of no food until new seedlings grow). People have relentlessly cut the forests and driven pandas into ever-smaller natural areas in this land of more than a billion human inhabitants. Chinese authorities, pushed by world opinion and their own affection for pandas, have responded admirably, but ever so late. The giant panda will probably survive, marginally in its few natural reserves, more dependably in zoos.

We will probably save most of the large species that interest or amuse us (we will lose—are losing at an accelerating pace—untold numbers of smaller, unnoted creatures). But salvation will not be in nature. Zoos are changing their function from institutions of capture and display to havens of preservation and propagation. We may applaud this revolution in concept, and we rejoice in the success of so many breeding programs. Yet the near certainty that most conspicuous species—like the panda—will survive only under human management fills me with sadness. Some of the reasons are practical—the problems of inbreeding, the disappearance of geographic variation as a subject for evolutionary study. But the primary reason is deeper, and hard to express. "Natural" and "artificial" represent a dichotomy not easily breached in human attitudes. An animal outside its appropriate historical place loses more than a home. When the Shunammite woman built a room for Elisha and furnished it with bed, table, stool, and candlestick, the holy man asked, "what *is* to be done for thee? wouldest thou be spoken for to the king?" (2 Kings 4:13). She replied, with beautiful conciseness, that she wanted nothing, for she lived in greatest satisfaction: "I dwell among mine own people."

On the subject of biblical metaphor, please do not forget that Elisha arranged for her to conceive a son nonetheless, and later raised him from the dead. Good luck to the panda.

2

Cardboard Darwinism

1.

Darwin began the *Origin of Species* not with fanfare, but with fantails—pigeons, that is. He wrote in Chapter 1:

> Believing that it is always best to study some special group, I have, after deliberation, taken up domestic pigeons . . . I have kept every breed which I could purchase or obtain . . . I have associated with several eminent fanciers, and have been permitted to join two of the London Pigeon Clubs.

The public often equates the best science with the biggest questions. Surely the heroes of science are those who dare to ask how the brain thinks and where the universe ends. Practicing scientists, though not unmindful of these deepest conundrums, know that such questions, however vital and thrilling, are vacuous (at least for now) if we have no data for testing competing hypotheses and don't even know where we might find the requisite information. Progress in science, paradoxically by the layman's criterion, often demands that we back away from cosmic questions of greatest scope (anyone with

A review of *Vaulting Ambition* by Philip Kitcher, *Myths of Gender* by Anne Fausto-Sterling, and *Females of the Species: Sex and Survival in the Animal Kingdom* by Bettyann Kevles.

half a brain can formulate "big" questions in his armchair, so why heap kudos on such a pleasant and pedestrian activity). Great scientists have an instinct for the fruitful and doable, particularly for smaller questions that lead on and eventually transform the grand issues from speculation to action. While Lamarck (though a great empiricist on other subjects) selected an armchair as the source for his evolutionary treatise, Darwin chose pigeons, and revolutionized human thinking. Great theories must sink a huge anchor in details.

Thomas Kuhn's seminal work, *The Structure of Scientific Revolutions,* affected working scientists as deeply as it moved those scholars who scrutinize what we do. Before Kuhn, most scientists followed the place-a-stone-in-the-bright-temple-of-knowledge tradition, and would have told you that they hoped, above all, to lay many of the bricks, perhaps even set the keystone, of truth's temple—the additive or meliorist model of scientific progress. Now most scientists of vision hope to foment revolution.

We are therefore awash in revolutions, most self-proclaimed. Few programs for transformation have been more overt, few pursued with clearer aims in conscious sequence, than human sociobiology in the form proposed by Edward O. Wilson. The goal was audacious, but simply stated: to achieve the greatest reform since Freud in our notion of human nature. The citadel must fall in stages to the battering ram of strict Darwinism. The first step appeared as a chapter on the promise for a unified Darwinian theory of behavior, placed as a finale in Wilson's great treatise *The Insect Societies* (1971). Then the general theory (1975), *Sociobiology, the New Synthesis* (an explicit manifesto of revolution for the cognoscenti, since we tradesmen of evolution call our own Darwinian orthodoxy, a legacy of perceived revolution during the 1930s and 1940s, "the modern synthesis"). As *Insect Societies* ended with a chapter on sociobiology, so *Sociobiology* ended with a chapter on Darwinian explanations of human behavior.

Sociobiology included a diagram with a clear martial metaphor; it showed the social sciences eviscerated and then absorbed—half subsumed into neurobiology (as we understand how the brain works), the rest into Darwinian evolutionary theory (as we elucidate the value of behaviors in the ultimate Darwinian game of passing genes to future generations). Human nature would fall to the Darwinian tide in two stages. First, sociobiology tried to solve the easier problem of human universals, including gender differences that hold across cultures—the subject of Wilson's first explicit book on human sociobiology, *On Human Nature* (1978). But a theory of universal behavior cannot provide a comprehensive account of human nature; we must also encompass the differences among cultures, and the astonishing speed and lability of cultural change. Prima facie, the stately pace of Darwinian (genetic) change seems a most unpromising source for locating the differences, but here's the rub (or the "vaulting ambition" of Kitcher's title): the greatest revision since Freud must be comprehensive. What good is a theory of human nature that cannot explain the differences among us? Thus the dubious final step had to be taken, and a theory of differences proposed (via feedback loops between genes and culture, for no one could attribute such speed in cultural change to genes alone). Lumsden and Wilson vested their final step in a deeply flawed and now discredited mathematical model in *Genes, Mind and Culture* (1981)[1] and, shorn of equations for the layman, in *Promethean Fire* (1983).

But human sociobiology must be the most peculiar of self-proclaimed revolutions in science. We usually reserve this label for new structures of ideas, fundamentally new ways of knowing (evolution versus creation by fiat, or quantum indeterminacy against the Newtonian machine). Human sociobiology, by contrast, only raided one field with the unmodified tools of another. Moreover—and here the precious irony—sociobiology wielded the most orthodox version of these tools at the same moment that its parent discipline, evolutionary theory, had begun to reassess the very principles invoked to fuel the revolution in human nature. Human sociobiology worked by an untenable extension of flawed (even broken)

[1]See essay 7.

tools into uncongenial territory. Wilson's martial metaphor may represent hubris, and be quite mistaken—but it was certainly apropos.

As a severe critic of human sociobiology from its inception, I clearly am not an impartial observer. Yet surely the equation of bland nonpartisanship with objectivity—a silly notion fostered by the worst traditions of television news reporting—must be rejected. We may scrutinize a known critic more carefully, but ultimately we must judge his arguments, not his autobiography.

The debate on human sociobiology has been particularly murky because a legitimate political theme (misuse of genetic arguments to support existing social arrangements as biologically grounded) has accompanied a more interesting and abstract debate about modes of explanation from the start, a wrangle abetted by early Wilsonian statements that, to say the least, played poorly in feminist circles:

> In hunter-gatherer societies, men hunt and women stay at home. This strong bias persists in most agricultural and industrial societies and, on that ground alone, appears to have a genetic origin. . . . My own guess is that the genetic bias is intense enough to cause a substantial division of labor even in the most free and most egalitarian of future societies. . . . Even with identical education and equal access to all professions, men are likely to continue to play a disproportionate role in political life, business and science.[2]

Philip Kitcher has, I think, properly located the legitimate political issue as the greater care that should be demanded of those who choose to speculate when the consequences of confusing fact and guess are severe:

> The true political problem with socially relevant science is that the grave consequences of error enforce the need for higher standards of evidence. In the case of pop sociobiology, commonly accepted standards are ignored.

[2]E.O. Wilson, *The New York Times Magazine* (October 12, 1975).

> The mistakes merely threaten to stifle the aspirations of millions.

The dismissal of critics as politically motivated may bear some force as rhetoric for a counterattack, but such a defense only cheapens a fascinating intellectual debate. We raise the political point because it cascades from poor science.

Philip Kitcher and I approach the criticism of human sociobiology very differently, though we share a basic attitude. I believe that the methods and arguments of this field rest upon a central flaw. I think that the flaw lies in exporting to the analysis of human behavior the apparatus of the strictest form of Darwinian orthodoxy. That orthodoxy locates all evolutionary mechanics in the struggle among organisms for reproductive success. In Darwin's world, the calculus of success is simple—winners pass more copies of their own genes into future generations, nothing else. The theory has no space for such concepts as "the good of the species" or "the harmony of ecosystems," except as epiphenomena of the primal struggle for personal reward. This central character of Darwinism explains why sociobiologists have taken altruism as the test case for their approach—for, prima facie (though realities are deeper than appearances), acts of altruism could not be selected in Darwin's world, since they harm individuals, whatever they may do for species. Thus, if the phenomenon that seems most contrary can be rendered as consistent—by arguing, for example, that altruistic acts may save enough relatives to pass on sufficient copies of the altruist's genes—the general theory achieves stunning support through resolution of paradox.

The central flaw in sociobiology results from this Darwinian premise: the behaviors that the theory purports to explain must be interpreted as adaptations of organisms. At a time when evolutionary theory rings, above all, with criticism of these very notions, revolutions based on selectionist orthodoxy seem curiously anachronistic. Exclusive focus on organisms has been challenged by a hierarchical theory that grants equal weight to selection acting upon other entities of the genealogical hierarchy—genes and species, for example. Moreover, strict adaptationism has faltered badly as better understanding of genetic and developmental architecture forces us to view the parts of organisms as integrated into systems constrained by history and rules of structure, not as a set of tools, each individually honed to benefit organisms in their immediate ecologies.

Kitcher, on the other hand, takes another tack, and he has nearly convinced me. In a crowded literature, *Vaulting Ambition* is unquestionably the best criticism of human sociobiology that I have read. Kitcher argues that human sociobiology has no rigorous central core, that it represents a set of indifferent and largely speculative studies loosely coordinated by a common commitment to cardboard Darwinism. As such, it is slippery and elusive. Human sociobiology cannot be judged as an entity because it lacks coherence. Therefore, any proper assessor must undertake a patient dissection, case by case, until at the end nothing concrete remains to underpin the Darwinian superstructure. Darwin could invoke one hundred breeds of pigeons to support his theory of selection; human sociobiology is left with a single, flightless (and extinct) species of that ilk—the dodo. Kitcher writes:

> If we could expose one error underlying all the faulty analyses of human social behavior, then it would not be necessary to proceed, as I have done, by examining example after example. Unfortunately, sociobiology is a motley. Not only is there no single monolithic theory to be scrutinized, but the individual Darwinian histories offered by pop sociobiologists may be flawed in any of a number of different ways. There is a family of mistakes, and in distinct examples distinct members are implicated.

As a philosopher of science with strong skills in mathematics and a good background in evolutionary theory (he has also written the best book on creationism, *Abusing Science*), Philip Kitcher is uniquely qualified for this dissection. I admire most of all—for such diligence is rare in our frenetic world—the simple patience that Kitcher has mustered to invest such attentive care in the intricate details of case after case. A number of critics, myself included, might have done so before, but we didn't have the gumption or the endurance once it dawned upon us that no shortcuts could be found. No critique is so damning as the sequential removal of examples, one after the other, for as Mies van der Rohe proclaimed in my absolute favorite among mottoes: God dwells in the details.

But Kitcher also uses the special tools of his profession. I much appreciated the traditional skill of a trained philosopher in defining and dissecting arguments—particularly in the last chapter where I finally learned to define properly what had bothered me intuitively about the false bridges constructed by sociobiologists between Darwinian imperatives and human ethics. More important, Kitcher commands the mathematical skills to expose the fallacies and false assumptions of the apparatus needed by Wilson to complete his putative revolution—a model to explain differences among societies as anchored in genes and their interaction with culture.

Kitcher's critique begins with two crucial distinctions—first between broad and narrow sociobiology. Clearly, there must be a potential evolutionary science of behavior. If we wish to call this enterprise sociobiology (broad version), then no right-thinking person can oppose it. Kitcher's critique centers upon the narrow version, described above and vested in the strict Darwinism of natural selection working on items of behavior for their benefit in increasing the reproductive success of individuals.

Kitcher correctly notes that the narrow theory has no separate intellectual status—it is no more nor less than ordinary strict Darwinism applied to items of behavior rather than the shapes of bones and the colors of wings:

> As we approach, the "new synthesis" turns out to be a mirage. There is no autonomous theory of the evolution of behavior. There is only the general theory of evolution. . . .

> The problems faced by the aspiring sociobiologist are just those of other evolutionary biologists. Writ large.

For this reason, I have long maintained that the failure of human sociobiology lies not so much in the intransigence of *Homo sapiens* as in the increasing inadequacy of strict adaptationism as a general approach to evolution, exacerbated by some special problems posed by our own species (see below).

The second distinction contrasts sober with pop sociobiology. Kitcher defines the pop version, the subject of his critique, as: "appealing to recent ideas about the evolution of animal behavior in order to advance grand claims about human nature and human social institutions." Pop sociobiology does not designate the derivative writing of journalists and science writers, in contrast to the research of academic scientists, for Kitcher takes no cheap shots and his targets are all prominent professional biologists. It would also be a great mistake to equate sober versus pop with studies of other animals versus speculations about humans. Pop sociobiology is a style of speculative adaptationism equally applicable to herons, hamadryas baboons, and human beings. The issue is not taxonomy, but methodology. Studies of "rape" in mallard ducks and "adultery" in mountain bluebirds can be as pop as sociobiological arguments about human aggression.

Nonetheless, most sober sociobiology (applauded by all evolutionary biologists) treats nonhuman animals, while the worst excesses of pop sociobiology afflict *Homo sapiens.* Humans are particularly subject to the pop variety (speculative, storytelling adaptationism) for many reasons, three most prominent: (1) We have so little data about a slow-breeding species that cannot be overtly manipulated for experimental purposes. (2) The nongenetic process of cultural evolution often mimics the results of Darwinian (genetic) adaptation. In other species, limited consciousness guarantees that good behavioral design be shaped by evolutionary forces. Humans may devise good solutions without waiting for new genetic propensities, and then teach these advantages to their kids and neighbors. Cultural evolution is not even a good analogue for biological evolution because it proceeds so much more rapidly and, especially, because it works by amalgamation and coalescence across lineages—the very topology precluded on the Darwinian tree of continuous divergence. Thus we cannot even use the sociobiological models as good metaphors or analogies. (3) Our inordinate interest in *Homo sapiens,* and the attention that any new speculation receives, goads us to behavior that we would not countenance in studies of other species.

Sociobiologists often accuse their critics of falling into the oldest of Western cultural traps—the desire to keep humans apart from nature and free from her mechanisms, but *Homo sapiens* is not the only, simply the most prominent, victim of pop sociobiology.

Kitcher proceeds relentlessly through the bestiary of human sociobiology, beginning with general premises and the apparatus of Darwinian theory, proceeding through examples of evolutionary universals in human behavior—aggression, incest avoidance, gender differences, for example—to later claims for explanations of cultural differences, and ending with a chapter on the biological ground of ethics. He identifies a standard chain of invalid inference as "Wilson's ladder" and then proceeds to extirpate the rungs, case by case.

The ladder has four rungs, each subsequent step a dubious ascent from the perch below. Step one holds that we can use the general method for analyzing adaptation to argue "that all members of a group G would maximize their fitness by exhibiting a form of behavior B in the typical environments encountered by members of G." Step two claims that "when we find B in (virtually) all members of G, we can conclude that B became prevalent and remains prevalent through natural selection." I maintain that the ascent from step one to step two is the crucial weak link in adaptationist argument. The connection cannot be asserted a priori (though it often holds, in fact, and can be affirmed by a panoply of methods devised by evolutionists for the experimental and inferential study of adaptation. Unfortunately these methods, often requiring manipulation of breeding and environment, usually cannot be applied to *Homo sapiens,* hence the particular dilemma of this weak link for human sociobiology).

The automatic ascent from step one to two is a logical fallacy central to the adaptationist program: the confusion of current utility with historical origin. The mechanic's seal of current approval does not imply past construction for contemporary usage. Unfortunately, we call both the state of good design and the process of its origin by the same name—adaptation.[3] The false equation of one with the other is, in my view, the Achilles heel of human sociobiology. Belief in a higher power is, no doubt, markedly functional, even adaptive in evolutionary terms, for many people, but we cannot conclude from this current belief that such a notion arose by natural selection directly *for* religious ardor, nor can we infer that such belief is a genetically grounded entity at all (another requirement for Darwinian argument).

Step three holds that "because selection can only act where there are genetic differences, we can conclude that there are genetic differences between the current members of G and their ancestors (and any occasional recent deviants) who failed to exhibit B." My example of religious ardor also illustrates the fallacy of this connection. Behavior that works need not have a specific genetic ground, yet Darwinism is a theory of genetic change. Step four is the basis for political unhappiness with human sociobiology—the false equation of genetic grounding (itself unproved) with inevitability or resistance to change (many defects of vision are 100 percent heritable and easily corrected by a pair of glasses): "Because there are these genetic differences and because the behavior is adaptive, we can show that it will be difficult to modify the behavior by altering the social environment"—see the quotation from Wilson on page 29 for an example of this final fallacy.

If Wilson's logic of argument follows a ladder, then his strategy for a revolutionary theory of human sociobiology mounts in a definite sequence as well—from the validation of strict Darwinian adaptationism as a proper approach to human behavior, to the explanation of universal behaviors, and finally

[3]For an analysis of this problem see S.J. Gould and E.S. Vrba, "Exaptation—a missing term in the science of form," *Paleobiology*, vol. 8 (1982), pp. 4–15.

to the elucidation of cultural differences. The critique of Wilson's ladder questions step one of this strategy, but how does the sociobiological program deliver on its two empirical claims?

The debate on universals has properly centered on claims for adaptively based and genetically grounded differences between the sexes (an aspect of human social organization that transcends cultural particulars, according to supporters of sociobiological theory). The expectation for such sociobiological differences arises from the basic premise that all organisms, as the core of their being, must pursue the Darwinian imperative of individual reproductive success. In most animals, the argument goes, males and females must play the game differently, following the dictates of their biological roles. A sperm, little more than genes with a delivery system, is cheap to make, and each fertilization puts half of you into an offspring without further trouble. Thus, males should win their Darwinian edge by impregnating as many females as possible, as often as they can. This state of maximal spread may be achieved along a wide variety of routes, from stealth to outright domination—hence the variety of male strategies all tuned to one effect.

Females, on the other hand, invest much more by putting sources of nutrition into more expensive eggs; in addition, in many creatures, females bear offspring within their bodies and nurture their newborns for long periods. Hence, female investment must be prolonged and costly. Females receive no Darwinian edge in promiscuity since no gain in reproductive success attends any copulation after fertilization (while males can sew seed with their wild oats ad infinitum). Hence, female adaptations veer from profligacy toward care in choosing the best and most helpful males to father their offspring. The sociopolitical line of the pop argument now leaps from the page: males are aggressive, assertive, promiscuous, overbearing; females are coy, discriminating, loyal, caring—and these differences are adaptive, Darwinian, genetic, proper, good, inevitable, unchangeable. . . .

2.

We ought, however, to ask a more basic question before we assess this string of adjectives. Are the differences true? Do human males and females exhibit the predicted disparities? For if we cannot detect such universal distinctions amid the diversity of human cultures, what good can pop sociobiology provide? Its theory proclaims such differences as a fundamental prediction.

Myths of Gender by Anne Fausto-Sterling is a fine contribution to the empirical literature on human gender differences. If her argument is correct, then this fundamental sociobiological theory for human universals has failed, leaving very little of substance for the putative revolution. *Myths of Gender* is a courageous book, for it centers its conclusion upon the most undervalued and systematically ignored of all data—so-called null results.

Few observers outside science (and not nearly enough researchers inside) recognize the severe effects of biased reporting. The problem is particularly acute, almost perverse, when scientists construct experiments to test for an expected effect. Confirmations are joyfully reported; negative results are usually begrudgingly admitted. But null results—the failure to find any effect in any direction—are usually viewed as an experiment gone awry. Meticulous scientists may publish such results, but they disappear forthwith from the secondary literature (and are almost never reported in the press). Most scientists probably don't publish such results at all—who has time to write up ambiguous and unexciting data? And besides, they rationalize, maybe next week we'll have time to do the experiment again and get better results. I call such nonreporting perverse because we cannot gauge its depth and extent. Therefore, we do not know the proper relative frequencies of most effects—a monumental problem in sciences of natural history, where nearly all theoretical claims are arguments about relative frequencies, not statements about exclusivity.

Over and over again in my career I have bashed my head against this wall of nonreporting. When Niles Eldridge and I proposed the theory of punctuated equilibrium in evolution[4] we did so to grant stasis in phylogenetic lineages the status of "worth reporting"—for stasis had previously been ignored as nonevidence of nonevolution, though all paleontologists knew its high relative frequency. The critique of adaptationism fights a similar silence. A former student of mine recently completed a study proving that color patterns of certain clam shells did not have the adaptive significance usually claimed. A leading journal rejected her paper with the comment: Why would you want to publish such nonresults?

The study of gender differences suffers the same disabling bias. Measured differences are prominently reported, usually with much fanfare and much attention from the press. We really have no idea how often such differences are not found, for we don't know what results are simply not published. But, at least, we might tabulate assiduously *all* reported studies, positive, negative, *and* null results. This difficult and ostensibly unrewarding task forms the core of Fausto-Sterling's book on claims for differences in styles of intelligence and hormonally mediated disparities in behavior between men and women. Over and over again, she finds a disturbing pattern in the literature on "standard" gender differences. When differences are detected, they are usually in the direction proclaimed (though small in effect), but the great majority of studies report no difference, and these have dropped from sight. (For example, the attribution of different cognitive styles to women as a result of less lateralized brains—that is, less specialization between the two cerebral hemispheres—ranks among the most popular of current theories. Some studies find a small difference in this direction; none reports more lateralized brains in women. But most experiments have detected no measurable difference in lateralization—and this is the dominant relative frequency that should be prominently reported.)

Moreover, measured differences often correlate well with

[4]Eldredge and Gould, "Punctuated equilibria: An alternative to phyletic gradualism," in: Schopf, T.J.M., ed., *Models in Paleobiology* (Freeman, Cooper & Co., 1972), pp. 82–115.
Gould and Eldredge, "Punctuated equilibria: The tempo and mode of evolution reconsidered," *Paleobiology*, vol. 3 (1977), pp. 115–151.

immediate cultural distinctions, leaving little space for the sociobiology of genetically grounded adaptation. For example, poorer spatial perception is frequently cited as a different cognitive style for women. Fausto-Sterling shows that such differences occur in societies that greatly restrict the autonomy and physical movement of girls during early upbringing, but have not been detected in cultures (including Eskimos) that grant equal freedom to boys and girls. I need hardly say that this pattern, if as general as Fausto-Sterling suspects, would remove any empirical reason for invoking sociobiological explanations for the central issue in the study of human universals. Few general theories can survive the collapse of their crucial case.

If the sociobiological argument for universals fails, the more questionable extension to cultural differences hardly has a prayer of success. Kitcher's most effective section—Chapter 10 on "the emperor's new equations"—dissects the mathematical models developed by Lumsden and Wilson to close their system and complete the revolution. These models begin by acknowledging the obvious—that cultural differences are too rapidly developed to explain as genetically grounded adaptations. To save the Darwinian core of human sociobiology, cultural differences must be recast as the product of interaction between genes and culture—"gene-culture coevolution" in Lumsden-Wilson terminology. Thus even the labile and superficial vagaries of cultural shifts may be rooted in genetic adaptation promoted to speeds unattainable in Darwin's world by positive feedback through cultural reinforcement.

This line of argument culminates in the "thousand-year rule" of Lumsden and Wilson—the claim that one millennium is short enough to fix such genetically grounded differences among cultures. Kitcher has waded through the obfuscation of their formalisms to show that the models can work only under assumptions so restricted and unrealistic that few or no real cultures could follow their dictates. In particular, people must change from one cultural practice to another only with extreme reluctance (for otherwise, the tie of gene to specific practice will be broken as obviously favorable practices spread like wildfire to all observant people, regardless of genotype). Furthermore, the difference in success of the two practices must be large (or else the change cannot occur in a mere millennium). But what difference so favorably large will be so resisted? Kitcher concludes:

> The thousand-year rule is a theorem of the theory of gene-culture coevolution, when it is applied to the evolution of hypothetical people of extraordinary stupidity.

One begins to think that the entire revolution of human sociobiology now lies bereft of more than cloaking equations; its empirical underclothes are also missing—and it's a hard, cold world out there.

Avid adaptationists (A.J. Cain in particular)[5] have charged that critics of sociobiology have sacrificed the known truth of adaptation's hegemony in nature in order to provide self-serving arguments for pursuing a purely political dislike of sociobiology. If I may indulge one paragraph of autobiography, I can only maintain that, in my case, the location of strict adaptationism as the central fallacy of contemporary Darwinism had three major roots, two preceding sociobiology. The first arose from seven years' composition of *Ontogeny and Phylogeny* (1977), and my growing respect for the great European structuralist literature on laws of form (dating to such seminal thinkers as Goethe and Geoffroy). The second developed from a series of technical articles, written with David Raup, Tom Schopf, Dan Simberloff, and Jack Sepkoski between 1973 and 1977, on ordered patterns in phylogeny that arise within purely random systems (but were previously attributed without question to Darwinian adaptation).[6] Sociobiology did pro-

[5]A.J. Cain, "Introduction to general discussion," *Proc. Roy. Soc. London,* volume 205 (1979), pp. 599–604.

[6]D.M. Raup *et al.,* "Stochastic models of phylogeny and the evolution of diversity," *Journal of Geology,* volume 81 (1973), pp. 525–42; D.M. Raup and S.J. Gould, "Stochastic simulation and evolution of morphology—towards a

vide the third—as I struggled to understand what seemed so wrong about a speculative literature that reached conclusions about people so out of whack with my concepts of reality. We must not trivialize an issue so central and important as adaptationism with the cardboard notion that only base motives could inspire any opposition.

In any case, the charge that we have fallen into the oldest cultural trap of walling off *Homo sapiens* from the rest of nature, denying to people the evident Darwinian truths that we hold as self-evident for other animals, cannot be sustained for a simple reason: we question strict and speculative adaptationism just as strongly when its advocates never breathe a word about human beings and apply their methodology only to the brute creation.

3.

Bettyann Kevles has written a lively book, marred only by an overarching commitment to adaptationism in the a priori mode. *Females of the Species* is a survey of how female animals in nature play the Darwinian game of struggle for personal reproductive success. It contains not a word about *Homo sapiens,* except to say that nature's diversity precludes any hope for identifying any particular human propensity as *the* dictate of biology.

Nonetheless, this work concentrates on demolishing a reverse fallacy—the long tradition, now thankfully fading (with a substantial push to oblivion from this book), for interpreting what female animals do in the light of supposed role models imposed by sexist societies upon human females. Males in nature had often been viewed as the sole contestants in Darwin's game—endlessly scrapping, sneaking, pawing, and roar-

---

nomothetic paleontology," *Systematic Zoology,* volume 23 (1974), pp. 305–322; S.J. Gould *et al.,* "The shape of evolution: a comparison of real and random clades," *Paleobiology,* volume 3 (1977), pp. 23–40.

ing for success in mating. Females, if considered at all, were often viewed as passive acceptors of male victory (usually, females were just left out, or imagined as sitting by the hearth with kids while the boys were out hunting).

Yet Darwin himself, in establishing the notion of sexual selection (the real subject of his mistitled treatise, *The Descent of Man*), identified two components of the process, one male centered, the other female centered—male competition and female choice. Females may play as active and pugnacious a role as males in assuring their Darwinian heritage—by only consorting with acceptable males, and vigorously refusing all other propositions (often by rough rejection, not only by coy passivity). Yet even though the general theory had always featured equality in determination by females, few students of behavior (almost all, until recently, white males) paid much attention. Instead, they unconsciously viewed female society in nature as a mimic of sexist limits imposed on women by surrounding human cultures. Thus, for example, Solly Zuckerman's famous study of baboon behavior at the Regent Park Zoo in London (done in the 1930s) detected a male pyramid with an "alpha" baboon dominant on top. He scarcely considered relations among females beyond the presumed subservience of a cluster about the alpha male. The females, in fact, have hierarchies as well, for equally compelling Darwinian reasons, but they remained below a projected line of vision.

Kevles surveys the animal kingdom to demonstrate by the most powerful tactic of all—an overwhelming list of documented examples—that females are as active as males in following the Darwinian imperative. Her concluding paragraph reads:

> No one today would seriously suggest that female animals are mere egg repositories waiting for something to happen. Just as we are coming increasingly to appreciate the diversity of female roles in human society, so we are coming to understand the variety in the behavior of female animals—and to recognize females, as, at the very least, co-equal players in the evolutionary game.

The central argument of *Females of the Species* presents one aspect to celebrate, another to criticize. We students of natural history stoutly maintain that the beauty and virtue of our profession lie in the bounteous diversity of surrounding life; all things that can happen manage somehow to occur. Yet too many of our treatises (popular and technical) abandon this richness for abstraction, and use examples only sparely or to illustrate particular points, rather than simply to explore the boundaries of that wondrous range. Kevles has paid proper homage to nature. She introduces her subject with a minimalist ten pages of necessary generality, and closes it with three pages of warning about dangers of false extrapolation. In between, her text is an unrelenting bestiary of examples from beetles to baboons—hundreds and hundreds of tales about female behavior, organized in sections on courtship, mating, motherhood, and sisterhood, and all aimed to reinforce the argument that females participate in the Darwinian struggle for reproductive success as actively and as assiduously as males, only differently. Anyone can wax eloquent about diversity as nature's theme; Kevles has sunk years of work into its documentation.

Speculative storytelling in the adaptationist mode has been the primary weapon from evolutionary theory used by sexists to keep women in a subservient place. We might combat this weapon in one of two ways, and I think that Kevles has chosen the wrong path. One might, as Kitcher does, identify the fallacies of adaptationist methodology and call for a clean break with this invalid style of argument. Or one might, as Kevles does here (and as Hrdy and others have pursued before), try to construct a "good" sociobiology that depicts the natural behavior of female animals in a more positive light.

Kevles is not unaware of objections to strict adaptationism, and she takes account in occasional warnings, as: "There is not necessarily a purpose for every current structure, or a functional reason for every contemporary gesture." But when we read her copious examples, we are in Pangloss's world. Everything done by or to females is right, for all natural forms and actions are molded by the exclusive source of evolutionary change—the Darwinian struggle among organisms for individual reproductive success. Kevles concludes, in her final summation before the epilogue: "All female behavior, like the actions of males of the species, is self-centered. Individuals use and abuse each other, depending on which kind of behavior will help them survive."

Adaptation is a powerful force, but its sway is not exclusive—and we both caricature the process and ignore a central theme in current revisions of Darwinian theory when we equate adequate evolutionary reconstructions with our ability to tell a story about optimal behavior in the absence of definite evidence, and only because theory prefers this mode of argument, Kevles seems to feel that an evolutionist's work is done when she has told a plausible story about adaptation, not (as I would maintain) when such a tale can be tested with definite evidence against reasonable alternatives.

To cite just a few examples among hundreds (most centered on the conventional strategy of showing that even the oddest and most improbable behavior, infanticide and cannibalism for example, can also be for the best):

(1) Kevles tells us that sex is "for" the production of variety in offspring to secure evolutionary futures in a changing world:

> Let us assume along with Darwin that sexual variation permits a rapid accommodation to change of a cataclysmic type (though this is by no means a universally accepted explanation). For example, if the climate suddenly cools dramatically, those individuals with genes for thick fur will survive.

But you cannot sink one of the liveliest debates in all evolutionary theory into a parenthesis.[7] Besides, Kevles's adaptationist explanation for sex cannot be right, unless we totally

[7]J. Maynard Smith, *The Evolution of Sex* (Cambridge: Cambridge Univ. Press, 1978); G.C. Williams, *Sex and Evolution* (Princeton: Princeton Univ. Press, 1975).

misunderstand the nature of causality. In Darwin's world, organisms can only be selected for immediate advantages, not for success in unknown futures. Asexual organisms produce clones of offspring, sexual creatures a variety. An invariant clone may succumb in toto, but some members of a varying array may squeak through a future stress. Thus survival of a family line in the face of catastrophe may be a consequence of sexual reproduction, but this argument cannot explain why sex evolved in the first place, or why it continues in balmy times. Kevles confuses results and causes, a standard fallacy of adaptationist argument.

(2) In many species, males "protect" their fertilizations by sequestering mates, or even plugging their genital orifices with a hard secretion. Kevles is sure that such behavior, since it exists, must also be good for females (how about the alternative that males struggle continually with females, and sometimes one or the other sex wins): "These females have, in effect, traded off a guarantee of safety and sustenance for their offspring in exchange for an end to 'shopping around' and freedom." Her evidence is simple existence of the phenomenon. Whatever is, is right: "Although some of these arrangements seem unduly complicated, they apparently suit the females of the species, since they have acquiesced."

(3) When males seem to lord it over females, we must, according to Kevles, probe behind appearance and interpret domination as a fair bargain, equally good for females. On mating plugs: "Those females that accept plugs seem to have evolved according to the rationale that it is a safer bet to retain the supplies of sperm they have rather than to keep copulating in the hope of finding fitter fathers for their offspring." On reabsorption of fetuses from a previous copulation in the presence of higher ranking males: "She trades away certain fertilization with the sperm of a previous male for the opportunity to start the process all over with a seemingly better chance that her offspring will survive."

(4) When cuckoo nestlings live parasitically in the nests of redstart warblers, they may kill the redstart babies, yet redstart parents continue to feed the cuckoos. One might think that cuckoos have simply out-maneuvered the redstarts, but Kevles

insists that this situation must be best for redstarts as well. If redstarts haven't evolved mechanisms of defense or even of recognition, then selectionist pressures for such retaliatory behaviors must not be high after all, whatever the apparent benefit. If it isn't there, it isn't needed. Kevles writes: "Why have redstarts not evolved their own safeguards against the cuckoo's intrusion? It may simply be that there are many more redstarts than cuckoos and the mortality rate is not high enough to have prompted the adaptation of defensive strategies." But how about the alternatives that defense would be a fine thing indeed, but that redstart neurology precludes it, or that the situation has developed too recently for a response to evolve?

(5) Cannibalism must be prudent eating for the evolutionary benefit of all, not signals gone awry. Thus when an infant's bones turn up in the feces of a mother gorilla, we must assume (without any direct evidence) that the baby was doomed anyway, and that a good meal is better than nurturing a misfit:

> When primatologist Dian Fossey discovered infant bones in the feces of a gorilla mother whose baby had disappeared, Fossey deduced that the infant must have been maimed or moribund, and thus what looks at first glance like "murder" becomes "mercy killing." With the moribund infant out of the way, the mother will soon stop lactating, start cycling, and be able to become pregnant again.

It is all for the best; a Darwinian cipher is removed, and mom can start the evolutionary game anew.

Interestingly, Kevles is willing to ascribe cannibalism in zoos and laboratories to unnatural stress ("these events are unusual and probably pathological"), but similar occurrences in nature can only be active adaptations. Thus, hunters have noted in some species of mammals that fathers may eat newborns after mothers are killed: "Without their mothers' milk and care, the infants were doomed. The males' action in killing and eating the cubs, be they the father or just a passerby, is seen simply as making the best of a bad situation." (How about the alternative that males will always feed unless females are near to suppress them?)

(6) If males help in raising babies, they must be able to ascertain that the offspring are truly their own, for otherwise they are wasting time in Darwin's world (how about the alternative that females sometimes win this round and fool some nonfathers into caring?): "How they ascertain paternity is largely unknown. But they must have some way of judging because males in so many species do help, and selectively."

(7) Kiwi mothers play no role in rearing after laying their egg. This absence must be for the best, even though we might think that further investment of female energies would be better. (But why does this situation pose a problem at all? Maybe this nonoptimal system works just fine.) "The gray-brown kiwi produces the largest egg relative to body weight of all birds, and perhaps that enormous yolk, good nutritious material, is all the mother can afford to contribute. For the male incubates the kiwi chick and rears it with no female help at all." And so the observation of male rearing becomes, ipso facto, evidence of a best solution.[8]

(8) On the Channel Islands, ring-billed gulls often form female-female pairs. Kevles suggests that this homosexual bonding may be an adaptation for successful rearing when a promiscuous male flits momentarily from his own mate to fertilize one of the females: "It may be that homosexual pairing is a strategy that allows those extra females to get fertilized by promiscuous males and then to incubate their young as a mated couple." How about the alternative that an instinct for pairing overcomes the Darwinian imperative for reproductive potential on islands, like these, with unbalanced female/male ratios?

One might say that such speculative fancies are harmless, however relentlessly pursued. *Dulce est desipere in loco* ("it's fun to fool around once in a while"), as Horace wrote. But storytelling in the adaptationist mode is no mere indulgence, toler-

[8]S.J. Gould, "Of Kiwi Eggs and the Liberty Bell," *Natural History* (November 1986).

able for its good cheer, whatever its shortcoming as an alternative to proper science. For the deep belief embodied in the method—the idea that evolution is a tale of sequential molding of parts to designs favored in local habitats—badly misrepresents the richness of evolutionary theory. Evolution is the quintessential science of history, and the hold of history lies exposed in myriad imperfections and compromises (pandas' thumbs) featured by all organisms as legacies of their different pasts—while Panglossian adaptationism makes history irrelevant. If we inhabit the best of all possible worlds, we need not probe beyond or behind current function. Moreover, an atomistic theory of adaptive optimality denies the structural integration of organisms by interpreting each part as a separate puzzle in good design. An evolutionary theory that downplays both history and structure denies the very two subjects that make biology so special and interesting.

More basically, the Panglossian vision wallows in the most serious logical error of historical reconstruction, the same error that sinks human sociobiology—the equation of current utility with historical origin, or the idea that we know why a structure evolved once we understand how it works now. This error has extensive ramifications well beyond the confines of evolutionary theory. It invalidates, for example, the so-called anthropic principle now preached by some physicists and cosmologists who do not understand the lessons of history. The strong version of this principle holds, roughly, that since human life fits so intricately well into a universe run by nature's laws (current utility), these laws must have arisen with our later appearance in mind (historical origin).

The fallacy of inferring historical origin from current utility is best expressed by noting that many, if not most, biological structures are co-opted from previous uses, not designed for current operations. Legs were fins; ear bones were jaw bones and jaw bones were gill-arch bones; incipient wings could not power flight, but may have served for thermoregulation. The same error undermines the central claim of pop sociobiology. The human brain became large by natural selection (who knows why, but presumably for good cause). Yet surely most "things" now done by our brains, and essential both to our cultures and our very survival, are epiphenomena of the computing power of this machine, not genetically grounded Darwinian entities crafted specifically by natural selection for their current function. Pop sociobiology must stake a claim for unraveling the origins and meaning of human thought and behavior. But if most behaviors are co-opted epiphenomena, not selected entities, then sociobiological explanations in the adaptationist mode cannot touch them.[9]

A great asymmetry (I am tempted to call it *the* great asymmetry) pervades evolutionary histories. I am willing to admit that harmful structures will be eliminated by natural selection if they cause enough distress. But the other flank of the argument is not symmetrical—it is not true that helpful structures must be built by natural selection. Many shapes and behaviors are fit for other reasons. The continued success of flying fishes, as George Williams once noted, absolutely depends upon their propensity for falling back into the water after they emerge. But no one in his right mind would argue that mass was constructed by natural selection to ensure a timely tumble. Charles Darwin, who was not a mindless functionalist, but who struggled mightily to grasp the laws of structure, wrote (in arguing that unfused skull bones, though essential in permitting the large heads of mammalian neonates to pass through the birth canal, cannot be adaptations designed for this role):

> The sutures in the skulls of young mammals have been advanced as a beautiful adaptation for aiding parturition, and no doubt they facilitate, or may be indispensable for this act; but as sutures occur in the skulls of young birds and reptiles, which have only to escape from a broken egg, we may infer that this structure has arisen from the laws of growth, and has been taken advantage of in the parturition of the higher animals.[10]

[9]See essay 7 for an elaboration of this argument.
[10]C. Darwin, *The Origin of Species* (1859), p. 197.

Cardboard Darwinism—the central belief of pop sociobiology and the anchor too easily adopted for Kevles's adaptationist stories—is a theory of pure functionalism that denies history and views organic structure as neutral before a molding environment. It is a reductionist, one-way theory about the grafting of information from environment upon organism through natural selection of good designs. We need a richer theory, a structural biology, that views evolution as an interaction of outside and inside, of environment and the structural rules for genetic and developmental architecture—rules set by the contingencies of history and physicochemical laws of the stuff itself. The downfall of pop sociobiology will be a small benefit of this richer theory; its chief joy will be the deep satisfaction of integration: environment and organism; function and structure; current operation and past history; the world outside passing through a boundary (whether the skin of an organism or the geographical edge of a species) into organic vitality within.

3

Misserving Memory

Samuel Butler, a master of acrimonious polemic, confronted Charles Darwin with the sorest of all scientific subjects—a dispute about priority. In *Evolution Old and New* (1879), Butler accused Darwin of slighting the evolutionary speculations of Buffon, Lamarck, and his own grandfather Erasmus. To this book, and to later, more specific taunts,[1]

A review of *Darwin and the Mysterious Mr. X: New Light on the Evolutionists* by Loren Eiseley.

[1]Darwin arranged for the translation into English of a long German article by Ernest Krause on Erasmus Darwin. Darwin also wrote a preface to the translation. The extent of Butler's dedication to his anti-Darwinian crusade can be sensed in the fact that he learned German specifically to compare the translation with Krause's original! Indeed, Butler found several passages in the translation that had not appeared in Krause's article, and he felt that these had been added in order to confute his own deprecations of Charles Darwin in *Evolution Old and New.* Since Krause's original article had appeared before Butler's book, Butler charged that the added material seemed to condemn his own book "by anticipation," but had really been added *afterward* as a response to the book itself. He wrote to Darwin about this and Darwin replied: "Dr. Krause, soon after the appearance of his article in *Kosmos,* told me that he intended to publish it separately and to alter it considerably, and the altered MS was sent to Mr. Dallas for translation. This is so common a practice that it never occurred to me to state that the article had been modified; but now I much regret that I did not do so." But Butler was not satisfied, and he laid out his case in an excoriating letter in the *Athenaeum* for January 21, 1880. Darwin met him with silence and Butler then complained (in his book *Unconscious Memory,* published later in 1880) "that the occasion is, so far as I know,

Darwin reacted with the most effective ploy of all: silence. Darwin's son Francis later wrote of this incident: "The affair gave my father much pain, but the warm sympathy of those whose opinions he respected soon helped him to let it pass into a well-merited oblivion."[2]

Yet since the currency of science is originality in ideas, the subject of priority never passes into oblivion. Historians and sociologists have demonstrated over and over again that no revolutionary idea arises without a pedigree and that most major discoveries are, in the sociologist Robert K. Merton's term,[3] "multiples." (A.R. Wallace, of course, independently invented the principle of natural selection during a malarial fit on the island of Ternate eighteen years after Darwin had devised the argument, but before its publication.) The subject of Darwin's predecessors had been aired long before Butler's attacks. Most revolutionaries in science either ignore history or revise it for their own purposes. Darwin had not been generous toward previous evolutionists,[4] and their claims had led him to add a historical introduction to later editions of the *Origin of Species.*

---

unparalleled for the cynicism and audacity with which the wrong complained of was committed and persisted in." If ever there was a controversy worthy of the cliché about tempests and teapots, this is it.

[2]Francis Darwin, *The Life and Letters of Charles Darwin* (London: John Murray, 1887), vol. 3, p. 220.

[3]Robert K. Merton, "Resistance to the Systematic Study of Multiple Discoveries in Science," *European Journal of Sociology,* vol. 4 (1963), pp. 237–82.

[4]The reasons for Darwin's reluctance are complex and not all clear. In part, and legitimately, he did not wish to ally himself with what he regarded as the unsupportable, or even foolish, evolutionary theories of his predecessors. Lamarck had been ably refuted by Darwin's hero Lyell; while Chambers, author of the anonymous *Vestiges of the Natural History of Creation,* had been rightly ridiculed by many scientists, including Huxley, for such fanciful notions as the two-step transformation from bird to mammal through the intermediary of a duck-billed platypus. Darwin felt that he had to distinguish himself from other believers in the *fact* of evolution in order to gain support for his original *theory* of natural selection as its mechanism. Yet in part, Darwin did treasure the notion of priority, and did pursue it with perhaps too much zeal. He wrote to Hooker in 1859: "I have always had a strong feeling no one had better defend his own priority. I cannot say that I am as indifferent to the subject as I ought to be."

CHARLES DARWIN

Nor did the subject die with Butler's demise. In 1959, the centenary of the *Origin* inspired a flood of scholarship, much of it devoted to Darwin's evolutionary predecessors.[5] In the same year, Loren Eiseley, distinguished anthropologist and a notable prose stylist among scientists, published a long paper in the *Proceedings of the American Philosophical Society:* "Charles

[5]B. Glass, O. Temkin, and W.L. Straus, Jr., eds., *Forerunners of Darwin: 1745–1859* (Baltimore: Johns Hopkins University Press, 1959).

Darwin, Edward Blyth, and the Theory of Natural Selection." Eiseley charged that Blyth had developed a version of natural selection before Darwin in the 1830s and that "Darwin made unacknowledged use of Blyth's work."

Edward Blyth (1810–1873), the "mysterious Mr. X" of this book's title, was a gifted natural historian who, through misfortune and poverty, labored much of his life in relative obscurity as a curator in the museum of the Asiatic Society of Bengal. Eiseley, who died in 1977, did not pursue the subject much beyond his original paper, and this book (published in 1979) is little more than a reprint of the 1959 essay, fleshed out with seven additional essays, none original, many of doubtful relevance, and only one written later than 1965.

Two standard misinterpretations, one fostered by Darwin himself, have plagued conventional commentary on the genesis of natural selection. In the story that ought to be true, Darwin entered the *Beagle* laden with the theological prejudices of his age. During the next five years, nature herself cut Darwin's chains and he saw the truth in giant fossil bones from South America and in tortoises and finches on the Galapagos. This heroic tale reinforces the false stereotypes that scientists often like to convey about their discipline—science as a dispassionate search for objective truth through untrammeled observations; great scientists as people who can free their minds of prejudice and hear nature's own voice.

Darwin did begin to doubt aboard the *Beagle.* He did return with a prepared mind; and he did draw upon the resources of his voyage throughout his life's work. But he did not simply see evolution in the field. In fact, Darwin originally identified the famous Galapagos finches as members of several bird families. He put the story together only when a British Museum ornithologist later recognized the unity of design behind a striking diversity of adaptation.

But Darwin never claimed conversion through the *Beagle* voyage. In his own account,[6] source of the second misinterpre-

[6]See his *Autobiography* published in F. Darwin (note 2) and in several separate editions.

tation, Darwin opened his notebooks on the transmutation of species shortly after the voyage and groped for a theory to explain how evolution had occurred. (Darwin's demonstration of evolution as a fact must be distinguished from the theory he devised for its mechanism—natural selection. He accepted the fact before he developed his theory, and he never denied a long list of predecessors for the fact itself. He claimed originality only for the theory of natural selection.) Darwin did not underplay either his good preparation or the partial success of his first musings. But he does depict the genesis of natural selection as a "eureka" experience, a flood of insight after a lucky accident:

> In October 1838, that is, fifteen months after I had begun my systematic inquiry, I happened to read for amusement Malthus on *Population,* and being well prepared to appreciate the struggle for existence which everywhere goes on from long-continued observation of the habits of animals and plants, it at once struck me that under these circumstances favorable variations would tend to be preserved, and unfavorable ones to be destroyed. The result of this would be the formation of new species. Here, then, I had at last got a theory by which to work.

Eiseley's critique lies in this tradition of "eureka" interpretations. It accepts the idea of a watershed or Rubicon, a definite moment that propelled Darwin from groping uncertainty to satisfaction. But it transfers inspiration from Malthus to Blyth, and raises the favorite issue of all science gossip: did Darwin enhance the impression of his own originality by claiming tangential inspiration from Malthus rather than direct borrowing from Blyth? (Eiseley avoids the contentious issue of whether Darwin's slight was deliberate or unconscious.)

Twenty years have passed since the *Origin*'s centenary and Eiseley's paper, and no subject has been more diligently and fruitfully studied during this time than the source of natural selection. Through the work of Kohn, Limoges, Mayr, and, especially, Gruber and Schweber,[7] we now know that Darwin achieved no sudden triumph, but experienced more of a creeping conviction. Darwin's many notebooks have been reassembled and scrutinized minutely; his every coming and going during the weeks before he read Malthus have been reconstructed so far as the data allow. Gruber has proved that Darwin "groped" with far more purpose, direction, and success during the first year of his search. Schweber has traced the path of Darwin's month before Malthus and followed a trail from Comte, to Adam Smith and the Scottish economists, to Quetelet and the early statisticians, and from Quetelet's own citations, and not by serendipity, to Malthus. Darwin's "amusement," it seems, lay primarily in reading the Malthusian statement in its original formulation. He had previously studied the central claim that population must outrun supplies of food in a review of Quetelet's work.

Moreover, Schweber shows that most of the pieces were in place before Darwin experienced his Malthusian insight. His debt to Malthus may not extend beyond the quantitative formulation of a principle he already understood (and, in retrospect, an invalid quantification at that). This sensible reconstruction also resolves a major anomaly of the traditional "eureka" interpretation—why, if Malthus had removed opaque scales from his eyes, did Darwin not pause to celebrate, exult, or even record the event in his notebooks at the time itself? (Darwin wrote his autobiography in his old age, and largely as a moral homily for his children, not for publication. The famous Malthusian illumination cited above does not represent the only hazy recollection that led historians astray. Darwin's portrayal of himself as a dull, patient man, working on "strict Baconian principles," a fortunate fellow in

[7]E.D. Kohn, "Charles Darwin's Path to Natural Selection" (Ph.D. diss., University of Massachusetts, 1975); C. Limoges, *La sélection naturelle: étude sur la première constitution d'un concept* (Paris: Presses Universitaires de France, 1970); E. Mayr, "Evolution through natural selection: how Darwin discovered this highly unconventional theory," *American Scientist,* vol. 65 (1977), pp. 321–28; H.E. Gruber in Gruber and P.H. Barrett, *Darwin on Man* (Dutton, 1974); S.S. Schweber, "The origin of the *Origin* revisited," *Journal of the History of Biology,* vol. 10 (1977), pp. 229–316.

the right place at the right time, set an entire tradition of scholarship that denigrated Darwin until his intellectual resurrection of the last twenty years. Thus, we cannot view Darwin's misrepresentations in the *Autobiography* as consciously self-serving.) Incidentally, neither Gruber nor Schweber even mentions Blyth, and no major historian of Darwin has found a place for him in any of this recent work. Moreover, if there was no eureka, Blyth could only have been a minor actor in any case.

Even if later scholarship has swept the rug from Eiseley's style of explanation, I believe that his claim for Blyth falters on its own terms. In 1835 and 1837, Blyth published extensive articles in the *Magazine of Natural History,* a journal well known to Darwin. Blyth does indeed speak, in several passages, of nature removing individuals with disfavored traits. He admits that breeders employ a conscious version of this process to improve their stock, and he even wonders aloud whether nature might follow a similar course: "As man, by removing species from their appropriate haunts, superinduces changes on their physical constitution and adaptations, to what extent may not the same take place in wild nature, so that, in a few generations, distinctive characters may be acquired, such as are recognized as indicative of specific diversity?" Yet Blyth, as Eiseley admits of course, then quickly squelches his question with a resounding "no." "I would briefly despatch this interrogatory," he writes, primarily because intermediary forms cannot be found between most modern species.

Having rejected this suggestion, Blyth persistently discusses selection in an opposite sense—as a device for the removal of unfit individuals, leading to the preservation of species as constant and immutable, created entities. By acting as executioner for the sick, the weak, the deformed, the aberrant, the merely too large or too small, nature works with God to maintain species, not to change them. No notion of "natural selection" could be more contrary to Darwin's own (and, again of course, Eiseley acknowledges the difference). Blyth writes: "The original form of a species is unquestionably better adapted to its natural habits than any modification of that form." Selection, therefore, is a law "intended by Providence to keep up the typical qualities of a species." Predators impose constancy on their prey "by removing all that deviate from their normal or healthy condition, or which occur away from their proper and suitable locality, rather than those engaged in performing the office for which Providence designed them. . . . So profoundly wise are even the minor workings of the grand system; and thus do we perceive one of an endless multiplicity of causes which alike tend to limit the geographical range of species, and to maintain their pristine characters without blemish or decay to their remotest posterity." Not much solace for an evolutionist in that!

Eiseley's charge is thus reduced sharply in scope. He cannot claim that Darwin simply pinched the idea of selection from Blyth, for he must admit that Darwin's recasting of selection as an agent of *change* represents one of the great reformulations of thought in Western history. This creative topsy-turvy is the essence of Darwin's revolution; I cannot regard the possibility that he first encountered selection as a negative force in someone else's work as an issue of great import. But, even on this reduced charge, Eiseley cannot be sustained. For Eiseley was apparently unaware that selection, as a preserving force, had been a common concept of creationist biology. Blyth constructed nothing original; he was merely reporting general wisdom—selection as God's refining fire. C. Zirkle wrote a monograph on pre-Darwinian uses of natural selection as a guarantee that evolution could *not* occur.[8]

Since Darwin could have learned about selection as God's agent for constancy from a number of sources, Eiseley must prove more directly that he imbibed it from Blyth. Eiseley does indeed demonstrate that Darwin read Blyth's key papers, but so what? Darwin was a voracious reader. We know that he admired Blyth greatly and cited many of his specific observations in later work. (Blyth, by the way, returned both friendship and admiration, and never raised any intimation of unac-

[8]C. Zirkle, "Natural selection before the 'Origin of Species,' " *Proceedings of the American Philosophical Society,* vol. 84 (1941), pp. 71–123.

knowledged "borrowing.") Blyth's articles were published in a major journal that Darwin read regularly.

Eiseley's more specific claims for direct influence of Blyth's articles upon Darwin founder upon indefiniteness and inaccuracy. Eiseley begins his case, for example, with the fact that both Blyth and Darwin used the peculiar word "inosculate." Eiseley writes: "A rare and odd word not hitherto current in Darwin's vocabulary suddenly appears coincidentally with its use in the papers of Edward Blyth. . . . The rare and mildly archaic character of this word suggests that Darwin acquired it from his reading of Blyth." Yet, far from being archaic, the word "inosculate" was, at the time, a technical term in a very hot subject dear to the hearts of both Darwin and Blyth—the so-called "quinarian system" of taxonomy. The quinarians claimed that all organisms could be arranged into groups, depicted as five interpenetrating circles. They referred to the connections between circles as their inosculations (literally, their kissings). Blyth's article is, in fact, an explicit attack upon the quinarian system. Darwin's interest in barnacles (he wrote four volumes about them!) may be traced in part to their anomalous position within the quinary system, a scheme of classification that he liked no better than Blyth did.[9]

Yet Blyth and Darwin could not have been more different in their general vision of nature—and Blyth represented the old as firmly as Darwin pioneered the new. Their variant readings of natural selection represent the most striking expression of two incompatible views. To Darwin, selection is the creative force in evolution. If I had to summarize the essence of Darwinism in a single concept, I would emphasize the directing power of selection. Genetic variation is raw material; it is "random" in the sense that mutations do not arise preferentially directed toward the production of advantageous traits. Adaptation is the result of natural selection, acting relentlessly across generations to accumulate favored variation through the differential success of fitter individuals in producing more

[9]See my article, "The Rule of Five," in *The Flamingo's Smile.*

surviving offspring. Evolutionists have waxed poetic in their metaphorical depictions of selection—Ernst Mayr compared it to the work of a sculptor, George G. Simpson to a poet, Theodosius Dobzhansky to a composer, Julian Huxley to Shakespeare himself. The comparisons may be stretched, or even silly, but they do reflect the essence of Darwinism—the *creative* power of natural selection.

Blyth, on the other hand, was a natural theologian of the old school. Immutable unity of intended design was his watchword, and his arguments for divine beneficence were a bit outmoded even in his own time. He actually argued, in the crudest version of the argument from design, that God made mountains high enough to be covered with snow, so that the white color would reflect heat and prevent them from becoming "an increased source of cold by radiation to all around"; and that the moisture, thus retained in winter, would "furnish, by the action of the summer sun, a due supply of water, when needed, to the foundations and rills which irrigate and fertilize the more level country."

Blyth's world contained "one omnipotent and all-foreseeing Providence, under the beneficent dispensation of whom nought that ever exists or occurs stands isolated and alone, but all conduce and work admirably together for the benefit of the whole." The essential difference between old and new, between Blyth and Darwin, lies in Blyth's illustration of the divine unity: "It is the grand and beautiful, the sublime and comprehensive, system which pervades the universe . . . and which is so well exemplified in the adaptation of the ptarmigan to the mountain top, and the mountain top to the habits of the ptarmigan." Darwin's system permits us to speak of the ptarmigan adapting to the mountain top, for natural selection produces such apparent "order" in a world devoid of intrinsic purpose. But we can no longer speak of the mountain top adapting to the ptarmigan, and in this loss lies both the grandeur and the despair of "this view of life"—Darwin's own designation for his reconstructed world.

Disputes about priority, in any case, tend to ring hollow, despite the vigor of their persecution. Ideas are cheap. The use of ideas, the systematic reconstruction of a world in their light, is the stuff of intellectual revolutions. Patrick Matthew developed the notion of natural selection in 1831—in Darwin's later sense of a creative force. But who among you has ever heard of Matthew? He presented his position in scattered comments of an appendix to a book on naval timber and arboriculture—not the most auspicious location for a reconstituted world. The chief joy in reading Darwin lies in sensing the excitement of a man, not yet thirty, who knows that he holds a key to the reinterpretation of all biological and anthropological knowledge—and who pursues the reconstruction systematically. Matthew never saw the forest for his trees. Darwin cut through 2,000 years of philosophy with his statement in an early notebook: "Plato says in *Phaedo* that our 'imaginary ideas' arise from the pre-existence of the soul, are not derivable from experience—read monkeys for pre-existence."[10]

[10]My original review ended with two further paragraphs, too topical for inclusion here, that identify this book as both an editorial error and a poor way to honor the memory and accomplishments of so fine a man as Loren Eiseley.

4

The Ghost of Protagoras

The female mason wasp, *Monobia quadridens,* excavates a broad chamber by digging a long tube into the pith of trees and stems. She deposits a series of eggs in the tube, starting at the bottom and separating each egg from the next by a curved mud partition. The partitions are shaped with their rough and convex side toward daylight and their smooth and concave side toward the cul de sac at the blind end of the chamber. The larvae feed and pupate within their chambers, which the mother has provisioned with food. When the young adults emerge, they crawl toward freedom by chewing through the rough, convex sides of the partitions. If the partitions are experimentally reversed, so that the rough and convex sides now point toward the cul de sac, the emerging adults cut their way into the stem, pile up at the blind end of the tube, and eventually die. Apparently, the mason wasp has evolved a rigidly programmed rule of behavior: cut through the rough and convex side of the partition. In nature, obedience to this rule always leads to daylight. If a human experimenter intervenes to reverse the partitions, the wasp cannot accommodate and digs to its own death, steadfastly obeying its unbreakable rule.

What wasps lack—and what human beings possess in unparalleled abundance—is the common theme of both books: flexi-

A review of *The Evolution of Culture in Animals* by John Tyler Bonner, and *Man, the Promising Primate* by Peter J. Wilson.

bility in behavioral response. Bonner defines culture as "the transfer of information by behavioral means," and structures his fascinating book as a survey of culture in the animal kingdom, marching up the venerable chain of being toward bigger brains, increasing behavioral complexity, and freedom from rigid genetic programs specifying "single response behaviors." Wilson identifies flexibility—that is, freedom from genetic programming of specific behaviors—as the key to our evolutionary promise; he traces the origins of human culture to the structural and nongenetic (but biologically based) rules that we follow in establishing systems of kinship.

Human flexibility has at least three complex and interrelated sources. First, we possess a brain much larger, in proper relation to the size of our bodies, than that of any other animal (except the bottle-nosed dolphin). More circuitry gives any computing machine a capacity for flexible response that increases (indeed explodes) at a far faster rate than the growth of its material substrate. A simple machine can handle tic-tac-toe; complex computers may soon be giving chess grand masters a run for it. The metaphor may be somewhat mixed, but it is an arresting thought nonetheless that our brains contain more information, in an engineer's technical sense, than all the DNA in our genes.

Second, we have evolved our massive brains largely by the evolutionary process of neoteny: the slowing down of developmental rates and the consequent retention to adulthood of traits that mark the *juvenile* stages of our ancestors. We retain the rapid fetal growth rate of neurons well beyond birth (when the brain of most mammals is nearly complete), and end our growth with the bulbous cranium and relatively large brain so characteristic of juvenile primates. Neoteny also slows down our maturation and gives us a long period of flexible childhood learning. I believe that the analogy between childhood wonder and adult creativity is good biology, not metaphor.

Third, as primates we belong to one of the few groups of mammals sufficiently unspecialized in bodily form to retain the morphological capacity for exploiting a broad range of environments and modes of life. A bat has committed its forelimbs to flight, a horse to running, and a whale to balancing and paddling. Culture and intelligence at a human level may have required the evolution of a free forelimb and a generalized hand endowed with the capacity to manufacture and manipulate tools (both from *manus* = hand). Only the morphologically unspecialized among mammals have not made inflexible commitments to particular modes of life that preclude this prerequisite for intelligence.

Despite their common emphasis on flexibility as the hallmark of humanity, Bonner and Wilson have written very different books from disparate intellectual traditions allied to their professional affiliations (Bonner is a Princeton biologist, and Wilson a New Zealand anthropologist). I must confess that, although I share Bonner's profession, I find myself more attracted to Wilson's position.

Ecologists and evolutionists complain bitterly that molecular biologists sometimes denigrate their subject by attempting to explain whole organisms in terms of the chemistry and physics of constituent parts. Evolutionists decry this crass reductionism and defend the validity of their enterprise with appeals to pluralism and theories of hierarchical levels. Yet, like the boy who swore, when young, that he would never kick little kids off the basketball court when he grew up, but then couldn't resist the power of maturity, evolutionists sometimes take as haughty an attitude toward the next level up the conventional ladder of disciplines: the human sciences. They decry the supposed atheoretical particularism of their anthropological colleagues and argue that all would be well if only the students of humanity regarded their subject as yet another animal and therefore yielded explanatory control to evolutionary biologists.

Bonner's book does embody this style of reductionism in a curious feature of its organization. Explicitly rejecting any discontinuity in nature as a violation of some preeminent principle of uniformity, Bonner admits to a personal necessity for discovering the true origins of human culture well before the evolutionary rise of humanity. He writes: "I am a uniformitarian and believe that all evolutionary changes were relatively gradual and that we can find the seeds of human culture in very early biological evolution." He therefore begins with the motility of bacteria (quick and flexible behavioral reaction versus slow genetic response) and, like Emerson's worm, "mounts through all the spires of form" toward human flexibility. He arranges the facts of animal behavior according to their rank in a chain of being defined largely by increasing size of the brain. Bonner writes:

> There is a direct inverse correlation with the time of appearance of a group in earth history and the size of its brain. At one end of the spectrum fish have small brains, while at the other end mammals have the largest. This suggests a trend toward increase in the ability to learn, toward an increase in the flexibility of the response.

Other criteria are also invoked, all loosely coordinated around the more-is-better principle. In one remarkable passage, Bonner ranks species of social insects by the size of their groups: "The reason for considering this a more advanced stage is that these colonies are larger, consisting of 300 to 80,000 individuals." Perhaps the biggest giveaway occurs in what may be a slip, where Bonner abandons his usual, sufficiently dubious terminology of "higher" and "lower" organisms, and speaks of prehuman creatures as "lesser animals."

If the chain of being were an ineluctable mode of organization, rather than (as I would maintain) an indefensible conceit, then Bonner's continuationist perspective could scarcely be gainsaid. Life would move in a single direction and humans, on top, would be properly analyzed in biological terms as the improved inheritors of all that came before. But evolution is a copiously branching network, not a ladder, and I do not see how we, the titular spokesmen for a few thousand mammalian species, can claim superiority over three quarters of a million species of insects who will surely outlive us all, not to mention the bacteria, who have shown remarkable staying power for more than three billion years.

Bonner does sense his difficulty when he has to admit the undoubted success of his lesser creatures:

> Even though mammals have been successful in this evolutionary trend, and primates the most successful of all, not all the other vertebrates, the fish, the amphibians, the reptiles, and the birds should become extinct. Selection, more especially in a complex environment, will find many successful solutions each of which correspond to the available ecological niches.

But perhaps the proper lesson of this observation simply affirms Darwin's aphorism: "never say higher or lower." And if we abandon the venerable chain of being, we lose the most promising frame for viewing human culture in biological terms as an extension, almost a necessary one, of longstanding evolutionary trends.

Bonner's account also rests upon another reductionist tradition of argument—the atomistic and adaptationist style of modern Neo-Darwinism, which tends to dissolve entire organisms into constituent parts and to argue that most parts owe their form to direct shaping by natural selection. In this perspective, the problem of cultural evolution lies in discovering the adaptive significance of each individual feature. Bonner largely accepts the extreme form of the argument popularized by Richard Dawkins in *The Selfish Gene,* [1] and attacked with eloquence by Sewall Wright,[2] a founder of population genetics, who, at age 95, has been fighting this particular battle for more than half a century. Dawkins goes a step beyond reduction to parts and views genes themselves as the focus of natural selection. Bodies become mere "survival machines," temporary homes for genes engaged in a more than metaphorical struggle to make more copies of themselves in future generations.

I find little defensible in this view. Selection cannot "see"

[1]Oxford University Press, 1976.

[2]Sewall Wright, "Genic and organismal selection," *Evolution,* Vol. 34 (1980), pp. 825–43.

genes, and can only work through the differential birth and death of organisms. Nearly all genes have multiple effects, many irrelevant to adaptation. Bodies are not an inventory of parts produced by individual genes, but integrated structures that cannot always be changed piecemeal by the dictates of selection. Organic forms are not an array of optimal adaptations to their immediate surroundings, but complex products of history, not always free to change in any direction that might "improve" them. When Bonner writes that "natural selection for optimal feeding is then presumed to be the cause of non-motility in all forms," I can't help suspecting that some plants might do even better if they could walk from shade to sun—but that inherited constraints of design never permitted a trial of this intriguing option.

This research tradition of adaptive explanation by parts, inherited from the Neo-Darwinian synthesis of evolutionary biology, defines the essence of human sociobiology. In applying this style of argument to humans, many sociobiologists fail to appreciate the interposition of nongenetic cultural transmission as a cause of behavior. They reify the human repertory of behaviors into "things" (aggression, xenophobia, homosexuality), posit selective advantages for each item (usually by telling a speculative story), and complete a circle of invalid inference by postulating genes, nurtured by natural selection, "for" each trait.

John Bonner is far too subtle a thinker to engage in such contentious nonsense. He wisely avoids, indeed opposes by his emphasis on human flexibility, any sociobiological explanation for specific human behaviors. But this very wisdom forces Bonner into a dilemma that prevents an otherwise charming book from becoming a masterpiece. For Bonner is committed to various forms of reductionism—to the continuationism of a gradually ascending chain of being linking humans to the rest of life, and to the sociobiological penchant for adaptive explanations based on a struggle among genes. But since he is wisely unwilling to extend the argument to details of human behavior, he is left with rather little to say about the biological basis of human culture—little, in fact, beyond the unexceptionable claim that we have inherited, through large brains, the capacity to develop complex culture, and that this capacity is highly adaptive. So what else is new?

Bonner's problem is the central dilemma of all human sociology. When sociobiology is injudicious and trades in speculative genetic arguments about specific human behaviors, it speaks nonsense. When it is judicious and implicates genetics only in setting the capacity for broad spectra of culturally conditioned behaviors, then it is not very enlightening. To me, such an irresolvable dilemma only indicates that this latest attempt to reduce the human sciences will have very limited utility.

At this point Bonner might reply, with some apparent justice, that I have been unfair to him. His book is about culture in "lower" animals, and doesn't even have an explicit chapter on humans. And yet the ghost of Protagoras pervades the entire work, for man is the measure of all things. The decision to array animal cultures as a chain of being casts humans as the inevitable referent for all. We stalk every page, often in silence. Read and enjoy this book as a deft and lucid account, filled with fascinating stories of natural history. But do not expect a revelation at the top.

Peter Wilson's approach is entirely different. He also begins with a discussion of human biology, emphasizing the flexibility that arises from large brains, neoteny, and unspecialized morphology. But he then eschews any form of smooth extrapolation and argues that, while biology provides a necessary foundation, human culture represents a level of our lives that cannot be rendered in biological terms.

This antireductionist perspective is often attacked as obfuscating, if not downright mystical, by those of opposite persuasion. Bonner, albeit gently, caricatures the argument for an independence of human culture by writing:

> It is a new condition that came into being as a result of the complexity of the mind of early man. To that extent the cultural anthropologist would consider it biological, but once it came into being, it took a life of its own, and

its new properties cannot be understood in terms of the level below. It is, so to speak, self-propelled and, like a soul, has become detached from its body.

But the notion of hierarchical levels that cannot be reduced, one to the next below, is no appeal to mysticism. A claim for the independence of human culture is not an argument for fundamental ineffability (like a soul), but only for the necessity of explaining culture with principles different from the laws of evolutionary biology. New levels require an addition of principles; they neither deny nor contradict the explanations appropriate for lower levels. The principles of aesthetics do not preclude a chemical analysis of pigments in the Mona Lisa—but only a fool would invoke such chemistry to explain the essence of the lady's appeal. In this sense, the notion of partially independent hierarchical levels of explanation strikes me as a statement of common sense, not mystery or philosophical mumbo-jumbo. I also regard hierarchy theory as an indispensable approach for the proper analysis of human culture.

We must, of course, be wary of the deepest cultural prejudice of all: our almost desperate desire to make humans special and superior among the animals of our earth. We must therefore rigidly scrutinize any proposal for human uniqueness. We must recognize the elements of continuity that exist between the social behavior of human and other animals; and we must note that some animals maintain rudimentary expressions of what we would call culture in humans. Still, we cannot stretch our ladder of extrapolation from amoebae to monkeys and up to people. We are not better, but we are different because change in the basic fabric of our social systems occurs, as it does in no other animal, by the nongenetic transmission of information across generations—in short, by culture. (Yes, some birds teach songs to their offspring, and songs may alter by cultural tradition over generations—but major change in the basic structure of animal societies requires genetic modification.)

Darwinian—and all biological—theories of evolutionary change are, fundamentally, propositions about modes of genetic modification. What else could they be? Human cultural evolution proceeds along paths outstandingly different from the ways of genetic change. Biologists believe that genetic change is primarily Darwinian—that is, it occurs via natural selection operating upon undirected variation. Human cultural evolution is Lamarckian—the useful discoveries of one generation are passed directly to offspring by writing, teaching, and so forth. Biological evolution is constantly diverging; once lineages become separate, they cannot amalgamate (except in producing new species by hybridization—a process that occurs very rarely in animals). Trees are correct topologies of biological evolution.

In human cultural evolution, on the other hand, transmission and anastomosis are rampant. Five minutes with a wheel, a snowshoe, a bobbin, or a bow and arrow may allow an artisan of one culture to capture a major achievement of another. Fruitful analogies may be drawn between biological and cultural evolution, but they remain analogies. The processes are different, even though human culture has a biological base. Cultural evolution needs laws of its own. This statement is neither a council of despair nor a dashing of hopes for intellectual coherence. It is merely an acknowledgment of the world's hierarchical structure and, I hope, an intellectual challenge in its own right.

As an outsider to his profession, I cannot judge Wilson's particular argument for tracing a major root of human culture to the origins of kinship. But I do note that his style of explanation conforms to the hierarchical model that I find indispensable for analyzing human culture. Wilson does not eschew biology. In fact, he begins with biology and invokes it as a foundation in two senses, general and specific. We are "the promising primate" because our biological flexibility permits us to enter (for the first time in life's history on earth) the rapid and accelerating world of Lamarckian cultural evolution. More specifically, we develop kinship (the basis of our social structures) by producing an intersection between two fundamental, biological relationships. The "primary bond" links mother and child; the "pair bond" unites adult male and female. Other primates base their social groupings on one or the other, but do not combine them in any tightly integrated way.

The pivotal, and only common, figure in both bonds is the adult female. The missing concept, while the bonds remain separate, is fatherhood. Mammalian mothers must nurture their children, and (obviously) males and females must unite to produce offspring. But fatherhood, defined as nurturance by males rather than mere stud service, does not represent a biological necessity. Wilson argues that the concept of fatherhood arose in creatures sufficiently brainy to abstract the general principle of "relationship" and therefore to conjoin primary and pair bonds. Once the abstract principle of intersecting two relationships to form a third is grasped, it may be extended further to produce complex social groupings based on kinship. Wilson locates the foundations of human social organization in this structural, nonbiological rule and writes:

> If it is possible to create a relationship by combining two others, then the principle of transitivity emerges, by which a link between two individuals may be transferred to a third. . . . If it is possible to create the relationship of father by conjoining the pair and primary bonds, then since sex and generation boundaries have been crossed, there are no barriers left. The relations between any two individuals can be continued indefinitely by linking through the common member(s). The extension of the primary bond and the lengthening of the pair bond lead to a junction that is the elementary basis for a generalized social organization typical of the human species. For it is only then that kinship is possible, and with kinship we have the most flexible, generalized, and adaptable principle of group organization in the primate order.

The argument is based on biology—both the existence of pair and primary bonds and the braininess to grasp their abstract connection. But the result—the notion of fatherhood and the extension of its principle to systems of kinship—is not, in itself, a Darwinian phenomenon (though it has consequences for survival—and culture feeds back upon biology). As Wilson writes: "It is a distinguishing feature of the human species that although individuals are naturally impelled toward living in company and it is necessary for them to do so, there is no universal instinctive form that this company will take."

Wilson's title is a conscious *double entendre.* Humans are promising because their flexibility opens so many options. But the promise also assumes a more specific function. A woman cannot always play both roles of partner and mother simultaneously. She must often hold one in temporary abeyance. To secure the momentarily neglected role, she must make a form of representation indicating her future attention—in short a promise. Recognition of a promise requires the brain power to make abstractions and to develop a notion of the future. Wilson views the structural necessity for promising as a possible impetus to consciousness.

I am not taken by all of Wilson's methodology. He works in the tradition of Hobbes or Rousseau, arguing by presupposition and logic, rather than by empirical contact with the world. I am enough of a hide-bound Anglo-Saxon pragmatist to wonder about the ultimate worth of this older, speculative tradition. Still, I regard Wilson's work as promising in a third sense, for it acknowledges the legitimate uniqueness of human culture and extends a welcoming hand to biology at the same time. It debars the imperialistic designs of some biologists upon the human sciences, while it affirms as necessary the insights from Darwin's world. True partnership without engulfment or ignorance; I see no other path to a satisfactory science of human culture.

II

Time and Geology

5

The Power of Narrative

To begin the second act of Gilbert and Sullivan's *Patience,* Lady Jane enters a bare set, seats herself before her cello, and in two verses bemoans the changes of increasing age. In the first, or conventional, account, she laments what she has lost with the years, but in the second (speaking mostly of weight) she reports a steady increase: "There will be too much of me in the coming by and by!" The humor of this song plays upon our onesided notion that anything old must become battered, worn, and increasingly bereft of information.

Scholars often make the same false assumption that contemporary cases must provide optimal data, while the records of scientific work steadily decrease in breadth and reliability as they grow older and older. We might therefore suppose that, to understand science, a historian or sociologist should study debates and discoveries now in the making. Yet a moment's thought about our technological age should expose the fallacy in such an idea. Our machines have generally rendered data more ephemeral, or simply unrecorded. The telephone is the greatest single enemy of scholarship; for what our intellectual forebears used to inscribe in ink now goes once over a wire into permanent oblivion (barring such good fortune mixed with ethical dubiety as a White House taping machine).

A review of *The Great Devonian Controversy: The Shaping of Scientific Knowledge among Gentlemanly Specialists* by Martin J.S. Rudwick.

Moreover, in losing the art of writing letters, many scientists have abandoned the written word in a great many previous applications, from diaries (now passed from fashion) to lab notebooks (now punched directly in "machine-readable" format). The present can be a verbal wasteland. Paradoxically, then, our most copious data should, like Lady Jane, occupy a comfortable middle age—old enough to avoid our modern technological debasement, and young enough to forestall the inevitable losses of time's destruction.

"The great Devonian controversy" occurred during the 1830s, an optimal decade probably unmatched for density of recorded detail. The controversy began with an apparently minor problem in dating the strata of Devonshire; it ended with a new view of the history of the earth. Martin Rudwick can usually trace the course of its enormously varied and complex changes on a daily basis; little more than conversations over ale and coffee, or bedtime thoughts before candle snuffing, are missing. But density of data makes no case for significance on the fallacious premise that "more is better." The subject must also be important and expansive. After a superficial first glance, most readers of good will and broad knowledge might dismiss *The Great Devonian Controversy* as being too much about too little. They would be making one of the biggest mistakes of their intellectual lives.

The geological time scale is a layer cake of odd names, learned by generations of grumbling students with mnemonics either too insipid or too salacious for publication: Cambrian, Ordovician, Silurian, Devonian. . . . Their ubiquity in all geological writing has led students to suspect that these names, like the rocks they represent, have been present from time immemorial *(et nunc, et semper, et in saecula saeculorum, amen).* In fact, the geological time scale was established in an amazingly fruitful burst of research during the first half of the nineteenth century. In 1800, scientists knew that the earth was ancient, but had devised no scheme for ordering events into an actual history. The primary criterion for unraveling that history—the sequence of unique events forming the complex history of life as recorded by fossils—had not been developed. Indeed, many

GEOLOGIC ERAS

Era	*Period*	*Epoch*	*Approximate number of years ago (millions of years)*
Cenozoic	Quaternary	Holocene Pleistocene	
	Tertiary	Pliocene Miocene Oligocene Eocene Paleocene	
			65
Mesozoic	Cretaceous Jurassic Triassic		
			225
Paleozoic	Permian Carboniferous Devonian Silurian Ordovician Cambrian		
			600
Precambrian			

fine scientists still denied that species could become extinct at all on an earth properly made by a benevolent deity. Geologists of 1800 confronted a situation not unlike the hypothetical, almost unthinkable dilemma that historians would face if they knew that modern cultures had antecedents recorded by artifacts, but did not know whether Cheops preceded Chartres or, indeed, whether any culture, however old and different, might not still survive in some uncharted region.

By 1850, history had been ordered in a consistent, worldwide sequence of recognizable, unrepeated events, defined by the ever-changing history of life, and recorded by a set of names accepted and used in the same way from New York to Moscow. This "establishment of history" was a great event in the annals of human thought, surely equal in importance to the more theoretical, and much lauded "discovery of time" by geologists of generations just preceding. Yet while we celebrate Galileo, Darwin, and Einstein, who beyond a coterie of professionals has ever heard of William Smith, Adam Sedgwick, and Roderick Impey Murchison, the architects of our geological time scale and, therefore, the builders of history? Why has their achievement—surely as important an accomplishment as any ever made in science—been so invisible?

We must resolve this vexatious question in order to appreciate *The Great Devonian Controversy,* for Rudwick's seminal book tells the story of how one major period of the earth's history was recognized and unraveled. That period, the Devonian, occupies the crucial time between 410 and 360 million years ago, when life flourished in the seas, and plants and vertebrates became abundant and diverse on land. We will not grasp the importance of this achievement if we follow a common impression and consider stratigraphic geology as "mere description," to be dismissed as "narrative" in the primitive mode of unquantified storytelling. *The Great Devonian Controversy* challenges us to understand natural history as a worthy style of science, equal in rigor and importance to the more visible activities of measurement and experiment that set the stereotype of science in the public image. The success of this book might therefore prompt a broadscale reassessment of science itself as a human activity. *The Great Devonian Controversy,* in its unassuming and highly technical format, could become one of our century's key documents in understanding science and its history.

Since the history of science is usually written by scholars who do not practice the art of doing science, they usually impose upon this greatest of human adventures a subtle emphasis on theories and ideas over practice. (I exempt Rudwick, who had a first career as a distinguished paleontologist before switching to the history of science.) The late eighteenth-century Scottish geologist James Hutton, for example, is usually praised as the instigator of modern geology because his rigidly cyclical theory of the earth established a basis for an immense span of time. But Hutton had precious little impact on the practice of geology; his name became an icon, but his theory remained on a periphery of speculation; the doers of geology largely ignored his contribution and went about their work.

Theory, of course, necessarily permeates everything we do. But theory may also be pushed below consciousness by groups of scientists who choose to view themselves as ardent recorders of nature's facts. Early in the nineteenth century, the founders of the Geological Society of London explicitly banned all theoretical discussion from their meetings, and dedicated themselves to what they called the "Baconian" (or purely factual) recording of history. They relegated Hutton and other theorists of past generations to the shelves of speculation, and pledged themselves to fieldwork—specifically to the inductive construction of a stratigraphic standard for history.

When I was younger, and understood science poorly, I bemoaned what I considered the paltry spirit of these men. How could they abandon the exciting ideas, the expansive vision, that motivated Buffon and Hutton, and dedicate themselves instead to finding out what lay on top of what in the rocks? I now appreciate the motives of these men who, after all, forged the geological time scale with their own eyes and hands. Of course they were too extreme (even disingenuous) in their impossible rejection of theory; of course they overreacted to what they interpreted as past excesses of speculation. But they understood the cardinal principle of all science—that the profession, as an art, dedicates itself above all to fruitful doing, not clever thinking; to claims that can be tested by actual research, not to exciting thoughts that inspire no activity.

Most younger rocks of Britain and the Continent had yielded with relative ease to consistent stratigraphic ordering—for they are arranged nearly as an ideal layer cake, younger above older, with relatively little distortion by folding, breakage by faulting, or inversion when older layers thrust over younger strata. But older rocks, from the Paleozoic Era of our modern time scale (600 million to 225 million years ago), posed greater problems for three major reasons well known to all field geologists: rocks of this age are usually far more contorted by intense folding and faulting; alteration by heat and pressure has obliterated fossils from many units, thus removing the primary criterion of history; large stretches of time lie unrecorded by strata, because repeated episodes of mountain building and continental collision destroy evidence.

The complexity of Paleozoic rocks seriously threatened the stratigraphical research program. If older strata could not be ordered by resolving their structure through mapping, and by sorting their sequence according to fossils, then the science itself was doomed. This ordering, moreover, had great economic importance. The period of great fossil forests or "Coal Measures" (the Carboniferous Period of our modern time scale dating from about 360 to 280 million years ago) lay near the top of this older pile. If geologists couldn't figure out what strata came before and after, vast sums of money would be wasted drilling through rocks misinterpreted as young but actually older than Coal Measures, in the vain hope of finding coal beneath.

The rocks of Devonshire became a focus for resolving the stratigraphy of older times. Most strata were dark, dense, compacted, and highly contorted—composed of rock known as "greywacke" to quarrymen. Under the lingering tendency to infer age from rock type (an old hope that had failed the test in detail, but had not been abandoned as a rough guide), greywacke smelled old and seemed destined for a position at the bottom of the stratigraphic pile.

The Devonian controversy began in 1834 with a paradoxical claim about fossils supposedly found within the greywacke sequence of Devonshire. Henry De la Beche, later director of the Geological Survey of Great Britain, claimed the discovery of fossil plants *within* the Devonshire greywackes. (Don't let the Francophonic name fool you; De la Beche, né Beach, was a proper English gentleman, an important theme in Rudwick's analysis.) Roderick Impey Murchison, the aristocratic and wealthy ex-soldier who had parlayed a taste for outdoor adventure into a serious and fulltime professional commitment, responded in a manner that seemed shockingly inconsistent with the professed empiricism of the Geological Society.

Murchison had never seen the Devonshire rocks, but he proclaimed, with vigor and total assurance, that De la Beche had made a monumental mistake, a kind of fool's error in basic mapping and observation. The greywackes, Murchison argued, were old, belonging to the Silurian system (that he had named and first described himself); but the plant fossils must hail from Coal Measure (much later) times, since terrestrial life had not yet made an appearance during the Silurian. The plant-bearing strata, Murchison proclaimed, must lie on top of the stratigraphic pile, not within as De la Beche claimed. And the sequence must contain a prominent "unconformity," or temporal gap in deposition, if such young rocks (Coal Measures) actually lie directly atop the ancient greywackes.

(I present, with great regret and almost with a sense of shame, this simplified caricature of such a complex argument and its eventual resolution. As his primary theme, Rudwick emphasizes the multiplicity of forces and arguments leading to the resolution of such debates; and he urges us to view scientific change as a social process of complex interaction based on class, status, age, and place of residence. I have been forced, by limitations of space, to depict the debate largely as a struggle between two men and a pair of sharply contradictory interpretations later forged into unity. You simply must read the book to sample the true richness of the controversy.)

Murchison based his bold and cocky response on his confidence in fossils as an ultimate criterion for assessing the relative ages of rocks. No issue could be more important in developing a proper methodology for reconstructing history. Geologists had to find a reliable criterion—some feature of rocks or their contents that changed in an absolutely reliable way through time to produce a sequence of unique and unrepeated states for marking the measuring rod of history. Nature does not yield her secrets easily. Nowhere on earth can geologists find a complete pile of unaltered rocks—where relative age might be assessed by simply observing what lies on top of what (a procedure called "superposition" in geological jargon). Rock outcrops are fragmentary and jumbled; only a tiny part of the sequence graces any particular road-cut or stream bed. We need a criterion for linking these isolated pieces one to the other ("correlation" in the jargon). Just as archaeologists might use tree rings of support beams, styles of pottery, or forms of axheads to order a group of widely scattered pueblos into a temporal sequence, geologists needed, above all else, a criterion of history.

Geologists had toyed with a variety of criteria and rejected them. Rock type itself had been a favorite hope, the basis of the so-called Wernerian theory so popular in the generation just preceding the Devonian debate. (In Werner's system, all rocks precipitated from a universal ocean in order of density. Granite crystallized out first and more conventional sediments followed. The demonstration, by Hutton and others, that granitic cores of mountains had, in liquid form, intruded the overlying sediments from below, and did not form the base of a stratigraphic pile, turned Wernerian theory on its head—for the granite became younger than overlying sediment, not the oldest rock of all. Moreover, and more importantly, granite could be intruded at any time, and the mineralogical composition of a rock could not therefore specify its age.)

We can now understand Murchison's vehemence. William Smith in England (the surveyor and engineer adopted by the more patrician leaders of the Geological Society as their intellectual father) and Georges Cuvier in France had established fossils as a proper criterion of history. We now know, as Murchison and his colleagues did not, that evolution is the reason for the uniqueness of each fossil assemblage. Extinction is truly forever; once a group dies out, the hundred thousand unpredictable stages that led to its origin will never be repeated in precisely the same way. All genealogical systems of such complexity must share this property of generating unique sequences, and can therefore serve as criteria of history. But a scientist need not understand (or may incorrectly interpret) the causal basis of uniqueness. Its empirical demonstration will suffice.

By claiming that Coal Measure plant fossils could occur in Silurian or older rocks, De la Beche had made, in Murchison's view, the most retrograde possible step. He had abandoned the hard-fought and newly won fossil criterion of history, and had reverted to discredited rock type (the "old" appearance of the greywacke). The plants, Murchison claimed, were conclusive; the strata containing them must date from the Coal Measures.

The Great Devonian Controversy traces the resolution of De la Beche's and Murchison's diametrically opposite views into "a significant new piece of reliable knowledge about the natural world." Through four hundred pages of the most dense and complex documentation I have ever read, Rudwick shows that the eventual resolution was not one of those dull compromises that mix a bit of this with a little of that and end up squarely in a lifeless middle between extreme alternatives. The basic resolution introduced a new dimension to the argument, a type of solution that neither side could have foreseen, and that emerged from a swirling context of debate when both initial positions had reached impasses. As Rudwick concludes in an apt metaphor:

> The battle lines defended by [the two initial interpretations] having initially faced each other in opposition, filtered silently through each other, as it were, until they faced outward, leaving at their rear a domain defended by them both. . . . The development of a successful interpretation . . . resulted in claimed knowledge that was unforeseen, unexpected, and above all *novel.*

De la Beche soon admitted that he had mismapped the structure of Devonshire. The strata with the disputed plants did lie on top of the stratigraphic pile, not within it. But De la Beche stuck to his central claim that the plants came from old Silurian rock. He insisted that the sequence contained no unconformities (temporal breaks in sedimentation); the plants lay atop the pile, but the entire pile was Silurian. With greater reluctance, Murchison finally dropped his a priori insistence that an unconformity must separate old rocks from Coal Measure plants atop the stratigraphic pile. He admitted continuity of sedimentation, but continued to insist, by the fossil criterion, that the plants could not be Silurian. These modified positions were closer, but still irreconcilable, in a way even more so because agreement had been reached about basic data of structure and sequence.

The eventual resolution required a novel interpretation: the rocks containing the plants, indeed most of the strata of Devonshire, were neither old (Silurian) nor young (Coal Measures) but representatives of a previously unrecognized time in between (the Devonian Period). The fossil criterion had been vindicated—for Murchison could not know that plants of Coal Measure type had arisen in an earlier unknown age, while he had stated correctly and for certain that these plants could not have inhabited the well documented Silurian strata. The rock-type criterion had failed again—greywacke did not mean old. The Devonian concept proved enormously fruitful. It quickly achieved firm status as a distinct, worldwide fauna, recognizable in rocks from Russia to New York. By the early 1840s, Devonian had entered the geological lexicon.

As the subtitle of his book attests, Rudwick views scientific knowledge as a social construction, uninterpretable as nature speaking directly to us through bits of fact in a logic divorced from human context. It mattered intensely that the Devonian was codified in Britain by Anglican gentlemen who viewed the Geological Society of London as their intellectual home. Rudwick takes no sides in the silly and fruitless debate between realism and relativism. The study of social setting does not imply either the irrelevance or nonexistence of a factual world out there. A lot happened between 410 and 360 million years ago, and our geological record has entombed the events in copious strata. But a good deal of arbitrariness, or rather social play, must accompany the parsing of continuous time into a series of discrete, named periods. The rocks didn't have to be called Devonian. And the boundaries are not naturally fixed at 410 and 360 million years; the strata contain no golden spikes. These are largely social decisions. (In fact, the boundaries of the Devonian are poorly defined by the general standard that major divisions of the time scale should be set by episodes of mass extinction to form more or less natural units—for one of the five major extinctions of life's history occurs *within* the Devonian.)

These kinds of decisions are also vitally important. Scientists are apt to say—but don't think for a moment that they believe it—"Who cares what we call it so long as we know what happened." Anyone with the slightest understanding of reward in science knows in his bones that accepted names and terms define status and priority—for actual history is soon forgotten. (Charles Lyell surely pulled the greatest "fast one" in the history of science when he subsumed acceptable parts of both catastrophism—the doctrine of periodic destruction by cataclysms—and his own excessively gradualistic and ahistorical world view under his favored name "uniformitarianism," with victory as the father of geology as his reward.)[1] Murchison, a keen debater and a self-promoter if ever there was one, understood this reality acutely and campaigned assiduously for his cherished term, Devonian. "The perpetuity of a name affixed to any group of rocks through his original research," he wrote, "is the highest distinction to which any working geologist can aspire."

I know from my own experience as a participant in major scientific debates that the explicit record of publication is utterly hopeless as a source of insight about shifts, forays, and resolutions. As Peter Medawar and others have argued, scientific papers are polite or self-serving fictions in their statements about doing science; they are, at best, logical reconstructions after the fact, written under the conceit that fact and argument shape conclusions by their own inexorable demands of reason. Levels of interacting complexity, contradictory motives, thoughts that lie too deep for either tears or even self-recognition—all combine to shape this most complex style of human knowledge.

Just consider some—a pitifully small sample—of the levels that must be grasped in order to understand the Devonian

[1]See my book *Time's Arrow, Time's Cycle* (Cambridge: Harvard University Press, 1987).

controversy (note also that all but the first do not enter the public record). We might begin with the pressures of data, for nature does speak to us in muted tones. Second, consider the ideological setting, so often unmentioned in published papers that emphasize the fiction of empirical determination. Both Murchison's allegiance to the fossil criterion and De la Beche's reluctance to grant it pride of place set the outlines of the debate. Basic attitudes to history itself lie embedded in these differing commitments.

Third, the overall social setting of science: The chief participants in the Devonian controversy were not academic geologists (for the profession did not yet exist in this form), but gentlemen of private means (or genteel poverty that threatened their commitment and forced them to scramble in an age that had not devised the concept of federal grants). Their published papers were archival, for the arguments had long before been aired and modified in open, but semiprivate debate at the Geological Society and other venues. (Modern science, with its "invisible colleges," proceeds no differently.) Although nearly one hundred men (yes, as another social reality, all men) participated, only a dozen or so really mattered, and they effectively excluded (or milked for their assessed value) all others who were not of the right class, degree of commitment or expertise, or simply couldn't get to London for the meetings.

Fourth, consider practical utility. Murchison persuaded the Czar of Russia that he was wasting resources by drilling for coal in a basin of Silurian rocks. If De la Beche had been right about the Devonshire plants, his majesty might have excavated much warmth.

Fifth, the social and economic position of scientists: Murchison was well-off and free to travel (he was also busy campaigning for a baronetcy, which he eventually obtained). De la Beche's social status was high enough, but his family fortunes had fallen on hard times. He persuaded the government to set him up as head of a geological survey—with salary. Government commitments placed him at a great disadvantage to the peripatetic Murchison; often, he could not even attend meetings since survey work required his presence in the field. When

Murchison attacked his interpretation of Devonshire stratigraphy, De la Beche reacted with a fierceness hard to understand until you realize that a charge of incompetence in mapping truly threatened his livelihood and his ability to continue any geological work at all.

Sixth, differences in personality: Murchison, an aristocratic old soldier, was self-assured and curt almost to the point of egomania. He described all his activities in military metaphor. In his own words, he never did fieldwork simply to find out what happened, or had a friendly discussion with an adversary over a pipe and glass of wine; he waged campaigns. De la Beche, on the other hand, tended to modesty and self-effacement in his letters and public statements. He swore up and down, over and over again, that he didn't care who got the credit, so long as the truth became known. Taking them at face value, one could develop a strong fondness for De la Beche and an equally firm antipathy toward Murchison.

But here we must probe below the surface of private documents. In fact, both men behaved pretty much the same in their secretiveness and quest for status. De la Beche once privately sent some crucial rock samples to the British Museum, hoping to hide them from Murchison, for he could scarcely deny access if Murchison knew about them. Murchison found out and forced De la Beche's hand; De la Beche relented and swore (quite dishonestly) that he had meant no such thing. One ends up with a dubious feeling about De la Beche's two-facedness, and with a grudging admiration for Murchison's consistency, despite his brutal characterizations, as in this note to Adam Sedgwick:

> De la Beche is a dirty dog. . . . I know him to be a thorough jobber & a great intriguer & *we* have proved him to be thoroughly incompetent to carry on the survey. . . . *He writes in one style to you and in another to me.*

Rudwick somehow manages to encompass in his narrative all these influences, and all the swirling, sallying, trenching and retrenching of opinion in the debate itself. The book is organized as a very thick sandwich—60 pages on context and methods, 340 pages of documented narrative, and 60 pages of interpretation and conclusions. Bowing to the reality of harried lives, Rudwick recognizes that not everyone will read every word of the meaty second section; he even explicitly gives us permission to skip if we get "bogged down in the narrative." Readers absolutely must not do such a thing; it should be illegal. The publisher should lock up the last 60 pages, and deny access to anyone who doesn't pass a multiple-choice exam inserted into the book between parts two and three. The value of this book lies directly in its detailed narrative.

I read this book, particularly its dense narrative, with great joy, but not without criticism. In my view, *The Great Devonian Controversy* suffers from two kinds of problems linked to its greatest strength—the enormous complexity of its tale. This is not easy to say without sounding flip or even anti-intellectual, but the brain, like the eye, cannot focus on all depths simultaneously. One can lose important aspects of the general pattern by concentrating too strictly upon intricate details. The old cliché about trees and forests is hackneyed because it has merit. I cherish what Rudwick has done; it is a monument to scholarship in an age of mediocrity. But I do think that the years immersed in detail, indeed the love one develops for each tiny facet, have led Rudwick to weigh some of the nuances wrongly, and to miss a major message that the Devonian resolution offered to the history of geology.

My first criticism is structural. Rudwick faced a monumental problem in deciding how to convey his unprecedented detail, so brilliantly worked out, in comprehensible form. He chose, unfortunately I think, a procedure that virtually forecloses full understanding of his book to all but probably a few hundred people in the world. He decided, in short, to follow with uncompromising strictness the historian's proviso that past events should not be read in the light of later knowledge, perforce unavailable to participants at the time. I would never dispute this credo as a general statement; I have fulminated against "Whiggish" history with as much gusto as any man.

But one can make almost a fetish of a good principle and find that it has turned against you.

Rudwick has chosen to organize his story by strict chronology, and absolutely never to mention, never even to drop the slightest hint about, any development that occurred later in time. I do appreciate the rationale. In 1835, no one knew that, a few years hence, a new block of ancient time would resolve the controversy. The injection of foreshadowing can only distort our understanding of people's positions as they developed. Thus Rudwick never breathes a word about the Devonian solution until Buckland, Sedgwick, and Murchison think of it themselves.

The result is faithful to narrative, but deeply confusing. And why not? The participants themselves were incredibly confused; tell it from their point of view, and what else can you derive? Is this style of presentation best for modern readers? I could make head or tail of Rudwick's narrative only because, as a professional geologist, I knew the eventual resolution. I don't think that this knowledge (which I couldn't expunge in any case) distorted my perspective; rather, it gave me an anchor to mesh the disparate threads into a coherent tale.

To cite just one example where modern information would clarify rather than distort, Devonian rocks had not been unknown in Britain before the controversy. One of the most famous and prominent of British formations, the Old Red Sandstone (another of those lovely quarrymen's terms), contains a rich fauna of Devonian fishes. The Old Red had not been resolved for an interesting reason: fossils of marine organisms form the standard stratigraphic sequence, but rocks of the Old Red are freshwater in origin. Thus, the Old Red cannot easily be correlated with the standard sequence. A simple point, easily made in a sentence—and so enormously helpful in understanding the difficulties that the Old Red presented to geologists of the 1830s (who assumed that these fishes had lived in the sea). The Old Red and its problems circulate through nearly every page of Rudwick's narrative, but he never tells us about the freshwater solution because it wasn't devised until after the 1840s—and we are left confused about the Old Red even after Rudwick's story ends. Only in a short appendix, "The Devonian Modernized," do we finally learn the solution (but how many people will read the appendices?). Again, as a professional geologist, I knew the answer from the start—and could spare myself the frustration of repeating the contemporary confusion in my own mind. Mystery writers don't tell the end at the beginning, but even their most complex stories are orders of magnitude simpler than the Devonian controversy. You do need a scorecard, at least partially filled in, to tell the players of Rudwick's drama.

Yet I strongly defend Rudwick's narrative style, storytelling in the grandest mode. Narrative has fallen from fashion; even historians are supposed to ape the stereotype of physics and be quantitative, or cliometric. Fine in its place, but not as a fetish. Narrative remains an art and science of the highest order, but of different form. How fitting that a book defending the importance of those scientists who established geological history should also defend so ably the narrative style of historical writing itself.[2]

My second criticism is conceptual, for I disagree fundamentally with Rudwick's interpretation of the controversy's resolution. Rudwick's analysis is so fine-grained, his interest in every item so intense, that he does not provide criteria for ranking the theoretical importance of various issues that the controversy resolved. Rudwick sees, because he knows the details so well, that nobody "won" in the narrow sense of getting all the marbles. But how else are issues of truly great scope resolved? No one can possibly be right about everything the first time. When controversies pit first-rate scientists against one another, any resolution will take bits and pieces of all views. Who could be so misguided as to get everything absolutely wrong?

From this knowledge of intimate detail, Rudwick depicts the

[2]Rudwick's last chapter contains several wondrously complex diagrams, outlining the changing views and their resolution, and the roles of various actors in the drama. Some will read these charts as a cliometric excursion. They will misunderstand Rudwick's intent. The charts are not a quantification; they have no scale except the chronology of years. One cannot quantify the magnitude of a changed opinion. The charts are pictorial models of narrative arguments, brilliantly conceived as epitomes.

resolution as a grand compromise (with a novel solution, not a melding of original views). He even implies that social construction of anything this complex could scarcely, in principle, be judged by victories. He writes, for example: "The finally consensual Devonian interpretation could be regarded as the analogue of a successfully negotiated treaty, precisely because it incorporated the nonnegotiable positions of both sides and found a reconciliation between them."

I disagree. I read the Devonian resolution as a clear victory for Roderick Impey Murchison—as sweet and unalloyed as any I know in science. Rudwick's own data reinforce this claim. De la Beche largely drops out of his story about half way through. Murchison moves on to honor and recognition. He is hailed as the "King of Siluria"; he receives medals from the Czar of all the Russias; his Devonian (name and concept) becomes a significant period of the earth's history. He views his own "campaign" as victorious. His triumph, moreover, is no cynical result of his assiduous self-promotion. He wins because his concepts have succeeded as the foundation of historical reconstruction.

Murchison had begun with an outrageously audacious claim; he identified De la Beche's errors before he had seen any of the rocks. Of course Murchison was wrong about the Coal Measure age of the plants. But how could he have known that such plants also lived just before, in a time previously unrecognized? He did know that such plants, by the fossil criterion of history, could not be Silurian—and he was right.

When we step back far enough, we can view the Devonian debate as part of a larger struggle about proper criteria for the reconstruction of history. From this perspective, we can rank the relative importance of issues. The defense of the fossil criterion was paramount. Murchison said so again and again and directed his choicest invective at De la Beche's willingness to abandon this criterion for the convenience of resolving some rocks in Devonshire. Murchison was right, and his rightness remains the cornerstone of stratigraphic geology. It also permits us to understand the meaning of history.

Of course, the fossil criterion did not survive in the simplistic form that Murchison had advocated at the beginning—as I said, you never get everything right the first time. The correlation of Old Red fish with marine Devonian proved that different environments could sport different faunas at the same time—a point that De la Beche had urged. But this represents a refinement of the criterion, not a retreat. What you cannot have—or the criterion fails—is a wholesale extension, as De la Beche sought, of the same fauna through time; that is, you can find different faunas at the same time, but not the same fauna through greatly different times. History is a series of irreversible changes yielding a series of unique states.

I am not happy in the role of Miniver Cheevy. I have lived through the revolution of plate tectonics and know how exciting ment in geological theory feels. But some victories of knowledge are so central, so sweet, that once attained, we can never experience their like again. Could any intellectual thrill be greater than the codification of a proper criterion for history itself, and the subsequent discovery of a great chunk of distinctive time, valid and recognizable on a worldwide basis?

We must recover this commitment to history from the conventional and prejudicial ordering of sciences by status that dismisses the sequencing of events in time as mere narrative or description. What better way than Rudwick's elegant double achievement—an explicit defense of the wisdom and modernity of historical inquiry (narrative as a term of pride) applied to the codification of geological history as one of the greatest triumphs in human understanding.

of time (or "deep time" as McPhee calls it). I well remember the catechism I learned in grade school: Mt. Lassen, which erupted in 1914, is the only active volcano in the United States (Alaska and Hawaii were still colonial possessions at the time). Written history recorded no eruptions, so we declared that the internal fires had been quenched irrevocably. Lassen is but one peak in the Cascade chain; St. Helens is another. Every one of those volcanoes is potentially active—Shasta, Hood, Rainier, all of them. Mt. Rainier may bury Seattle before another earthquake levels San Francisco.

Geology presents its irreducible beauty in raw appearances—and who would gainsay this aesthetic component. But another, perhaps deeper, beauty lies in understanding. The Cascade volcanoes extend for hundreds of miles in a linear belt. Why are they so aligned? And why do they stop in northern California? Why do volcanoes tend to come in linear arrays anyway? Until the theory of plate tectonics revolutionized geology by constructing a new earth, these questions had no adequate answers. But now we recognize that dense oceanic rock is pushed downward (subducted) beneath lighter continental rock when two plates—the large, thin "wafers" that form the earth's upper layer—push into each other, one with oceanic rock and the other with continental rock at its margin. As oceanic rock slides into the earth below a continent, friction induces partial melting of the sinking plate and, perhaps, some of the surrounding mantle rock as well. This molten material may rise to form a chain of volcanoes on the earth's surface right above the sinking edge of the oceanic plate.

California struck (and either consumed or removed) a ridge on the ocean floor many millions of years ago. Hence, from San Francisco to the south, oceanic rock is no longer spreading into and under America; rather the Pacific is largely moving north along the San Andreas Fault. But a shortened ridge persists in the eastern Pacific north of San Francisco, and the oceanic rock on its eastern flank is still spreading under America, where some melts and rises to form the Cascade volcanoes. When I viewed the lineup—Rainier, St. Helens, and, looking out the opposite window, Mt. Hood—I could almost see the old floor of the Pacific Ocean a hundred miles below me, and I apprehended those majestic volcanoes as mere pimples, minor vents for a restless earth.

John McPhee has captured this duality—beauty in particulars and satisfaction in general understanding—in his superb, highly personal book of reflections. He recognized that plate tectonics had revolutionized geology during the last two decades and he wanted to understand how the profession (and its earth) differed from the one he studied in a college course taken before the great instauration. He knew that he could not obtain the information he sought from didactic sermons in textbooks or from interviews with leading figures in academic offices. He had to go into the field with geologists, and spend months observing daily scientific life, to grasp how a scientific revolution permeates the core of practice. For God does dwell in the details, and visceral comprehension can only arise from an immersion in particulars. As a scientist, I thank McPhee for understanding this central aspect of our work, for so few who write about us do.

Before plate tectonics, there was no discernible God at all; only the details. Traditional introductory geology courses were catalogues of names for time slots, rocks, and landscapes, and they deserved the label that generations of undergraduates bestowed upon them—rocks for jocks. Both McPhee and I started with a course like that. He gives his list of fondly remembered terms; I could supply mine. My favorite was "paternoster lakes"—the chains of connected pools that often form near the edges of glaciers and once reminded someone of beads on a rosary, hence "ourfather" lakes. The terms did provide an interesting name for my friend's pet rabbit—Inselberg Bornhardt. Beyond that I could never see much use for them. In the hands of a poet with good slides, such a course could be inspiring; usually it was insufferably dull. No matter though; it cannot be taught anymore. (And thank goodness for that; for I am not a poet, and I now teach the successor to rocks for jocks at Harvard.)

Plate tectonics has given us a unified theory for the behavior and history of the earth as a whole. It has coordinated and brought under a single rubric such disparate phenomena as:

the origin of Iceland and Hawaii, the coincidence of earthquakes and volcanoes with ridges and subduction zones, the youth of the sea floor and the greater age of continental cores, and a background for the great Permian extinction that wiped out up to 90 percent of marine invertebrates some 225 million years ago. (Iceland and Hawaii sit above vents for molten rock rising from the earth's interior. Ridges and subduction zones mark lines of activity where plates meet, spread apart, collide, and rub against each other. New sea floor forms at ridges, spreads out, and disappears into subduction zones on a cycle measured in tens of millions of years, while continents stand high on their plates and cannot be subducted. The great extinction coincides with the fusion of all continents into a single Pangaea.) These phenomena, and hundreds of others, are no longer isolated facts. We can now present the earth as an integrated unit. Our science—and our courses—have become exciting; geology has regained a prestige it has not enjoyed since we discovered deep time and discernible history in the early nineteenth century, and Europe's finest scientists became geologists.

As McPhee's book opens, he is, of all places, on the George Washington Bridge. It makes sense. The Atlantic Ocean has had an off-again-on-again history during the last half billion years. It existed at the dawn of modern life in the great Cambrian explosion nearly 600 million years ago, but closed gradually during the next 350 million years and disappeared entirely when New York collided with Europe at the end of the Permian. But the Atlantic opened again soon thereafter and, beginning as a puddle, and then a channelway, grew as the sea floor spread beneath it at the mid-Atlantic Ridge. Our modern continents carry some curious scars of the collision; for the breakup of Pangaea did not occur precisely along the lines of previous suturing. Bits of ancient Europe remain stuck to the extreme eastern border of North America (as we know from strictly European fossils found in their rocks), while smidgens of old America grace the western coasts of Norway and Scotland.

Much of New Jersey's geology records the stretching and

splitting that reformed the Atlantic. But New Jersey is a fossil; the action is now more than a thousand miles east at the mid-Atlantic Ridge. McPhee wanted to see plate tectonics at work, so he headed west with his mentor Ken Deffeyes, professor of geology at Princeton, to the Basin and Range province of Nevada.

The Basin and Range is a vast expanse of alternating parallel chains: hills and valley, hills and valley. This topography is the superficial expression of geological activity beneath. The earth's crust is being stretched in Nevada. In response, it has fractured in a series of faults running perpendicular (north-south) to the direction of stretching (east-west). At each fault, the rocks on one side fall while those on the other rise, producing an alternating series of hills and valleys.

The Cascade volcanoes exist because the East Pacific spreading zone still operates north of San Francisco. But, as I mentioned before, California overrode the same zone farther south, as the North American plate moved west, pushed by sea floor forming at the mid-Atlantic Ridge more than 4,000 miles to the east. Many theories have been advanced to explain the Basin and Range province in terms of plate tectonics, but all involve the fate of this overridden Pacific spreading zone. In one version, the entire zone was subducted beneath California and still operates under the Basin and Range, uplifting Nevada and tearing it apart. In any case, Basin and Range topography records the early history of a potential ocean. The parallel hills and valleys of Nevada may be a living reminder of eastern North America's visage just before the Atlantic puddle reappeared some 200 million years ago. McPhee went to the Basin and Range to see plate tectonics in action, and his title is a symbol of our reconstructed, restless earth.

Basin and Range is a series of disparate but organically connected chapters recording McPhee's personal journey. Some are loosely sequential, as McPhee travels with geologists from the fossilized fracture zones of New Jersey to the active topography of Nevada. Others explore the consequences of plate tectonics in different ways—by examining the practicalities of Deffeyes's search for silver in abandoned mine tailings, and by contrasting plate tectonics with another great reconstruction of the earth presented by James Hutton in the late eighteenth century.

Where McPhee's style works—and it usually does—he triumphs by succinct prose, by his uncanny ability to capture the essence of a complex issue, or an arcane trade secret, in a well-turned phrase. Consider his topographic characterization of the United States: "really a quartering of a continent, a drawer in North America. Pull it out and prairie dogs would spill off one side, alligators off the other. . . ." Or his organic metaphor for "deep time": "With your arms spread wide . . . to represent all time on earth, look at one hand with its line of life. The Cambrian begins in the wrist, and the Permian extinction is at the outer end of the palm. All of the Cenozoic is in a fingerprint, and in a single stroke with a medium-grained nail file you could eradicate human history." If geology has two great themes—deep time and ceaseless motion—consider his one-liner for the second: "If by some fiat I had to restrict all this writing to one sentence, this is the one I would choose: The summit of Mt. Everest is marine limestone."

Yet the source of McPhee's greatest strength—his willingness to consort with scientists on a daily basis—is also the locus of his major weakness. He has, in short, imbibed too much of our mythology. In particular, he has been beguiled by the mystique of field work. No geologist worth anything is permanently bound to a desk or laboratory, but the charming notion that true science can only be based on unbiased observation of nature in the raw is mythology. Creative work, in geology and anywhere else, is interaction and synthesis: half-baked ideas from a barroom, rocks in the field, chains of thought from lonely walks, numbers squeezed from rocks in a laboratory, numbers from a calculator riveted to a desk, fancy equipment usually malfunctioning on expensive ships, cheap equipment in the human cranium, arguments before a roadcut.

This mythology leads to serious, but unfortunately conventional, misrepresentations of the past, a tradition that McPhee follows in the historical section of his work. (The rest of this essay is not an attack on McPhee's fine book, but a plea that we all reassess a common habit of thought.)

Ceaseless motion, one of geology's two great themes, achieved its explanation only when the theory of plate tectonics developed during the 1960s; McPhee's book concentrates upon this contemporary revolution. But deep time triumphed 150 years ago, and this first conceptual overhaul is best treated historically. The conventional hero for deep time is the Scottish gentleman farmer and geologist James Hutton (1726–1797); and while I do not dispute the attribution, I do challenge the usual explanation, uncritically adopted by McPhee, for Hutton's insights.

Hutton developed a theory of the earth that had, as its correlate and prerequisite, the proposition that "time, which measures everything in our idea, and is often deficient to our schemes, is to nature endless and as nothing."[2] He proposed an ever-cycling "world machine" that would make the earth's history as orderly and timeless as its celestial motion. Continental rocks are eroded and deposited in thick sequences of sediments at the bottom of oceans. Subterranean heat then fuses and consolidates the sediments to rock. The same heat expands, fractures, and uplifts the rock to new continents and the old, eroded continents become ocean basins. Land and sea change places, but the world always remains the same overall. The cycle then starts again. God once made a beginning and will undoubtedly ordain an end, but these are matters beyond the ken of science. As geologists, we can only observe and infer the cycles of the world machine; nothing in these investigations permits us to discern any direction, any indication that cycles will terminate in the future or ever differed in the past. Thus, in the last lines of his initial presentation of 1788, Hutton penned the most famous words in the history of geology: "The result, therefore, of our present enquiry is, that we find no vestige of a beginning—no prospect of an end."[3] Deep time with a vengeance.

In the conventional view, Hutton achieved his insight because he reasoned objectively from data observed in the field,

[2]James Hutton, *Theory of the Earth,* Transactions of the Royal Society of Edinburgh, vol. 1.
[3]Ibid., p. 304.

while his benighted opponents held fast to Moses from their pulpits or their academic desks. McPhee portrays Hutton as a field geologist, drawn to his theory of cycles inductively, after piling fact upon fact: "Wherever he had been, he had found himself drawn to riverbeds and cutbanks, ditches and borrow pits, coastal outcrops and upland cliffs. . . . He had become preoccupied with the operations of the earth, and he was beginning to discern a gradual and repetitive process measured out in dynamic cycles." He labels Hutton's concept of time as "novel and all but incomprehensible," and tends to view his opponents as religiously motivated and devoted to the strict Mosaic chronology of a mere 6,000 years or so for the earth's history.

Hutton did do some field work, and it did him a world of good. But his system arose as much from his culturally bound mind as from his beloved rocks. Hutton wasn't fighting theology; he merely acknowledged a different God—the perfect clockwinder who ordained proper laws of nature when he created the universe, and then let it run without further meddling.[4]

Hutton's system cannot be understood until we recognize its grounding in a view of science, now repudiated and even slightly musty in Hutton's own time: all phenomena and processes have not only an "efficient cause" (a mechanical explanation coincident with our entire sense of the word "cause"), but also a "final cause," or purpose. The old Aristotelian notion of final cause still has a place in discussing human conduct or the adaptations of organisms (though we now view the agents as natural rather than divine). But final cause has been banned from physical science; we may not speak of the purpose of erosion or lunar motion.

But Hutton was committed to the belief that all phenomena had both efficient and final causes: God wound the clock correctly and events unfolded in accordance with his preordained purpose. Hutton referred to his work as "this view of things,

[4]See my book *Time's Arrow, Time's Cycle* (Cambridge: Harvard University Press, 1987) for an expanded treatment of Hutton's views about time and history.

where ends and means are made the objects of attention." When we treat Hutton as a modernist and discuss only the mechanical workings of his world machine, we cannot understand him. Hutton repeatedly stated, both in private and in the organization of both his great treatises, that the world machine had its roots in a troubling challenge to *final* cause.

As a farmer by trade, Hutton was obsessed by what might be called the "paradox of the soil." Soil, as a substrate for agriculture is a *sine qua non* of our life on earth, and God made the earth for us: "This globe of the earth is a habitable world; and on its fitness for this purpose, our sense of wisdom in its formation must depend." Soil is a product of erosion and decomposition of rocks. But erosion also destroys the land and will eventually carry all sediments to the sea. Would God construct such a world where the same agent that supports life must eventually destroy it by wearing away the continents? Therefore, a restorative force *must* exist; sediments deposited in the sea must be raised to form new continents.

Hutton's work was placed in the category of old-fashioned, speculative world systems by the most committed empiricist among Hutton's immediate successors—the French catastrophist Georges Cuvier.[5] And who can blame him when we read the opening paragraph of Hutton's 1788 treatise:

> When we trace the parts of which this terrestrial system is composed, and when we view the general connection of those several parts, the whole presents a machine of peculiar construction [efficient cause] by which it is adapted to a certain end [final cause]. We perceive a fabric, erected in wisdom, to obtain a purpose worthy of the power that is apparent in the production of it.

In the simplistic scenario of hero=uniformitarian=empiricist vs. villain=catastrophist=theological apologist, Cuvier falls among the damned because his belief in rapid changes supposedly upheld an earth of limited antiquity, and God's

[5]See the section "of former systems of geology" in the famous *Discours préliminaire* of 1812.

direct role in geological history. In fact, none of the great catastrophists followed Moses, and their method was more rigidly empirical than that favored by any uniformitarian. They believed what they saw in the rocks—abundant evidence of catastrophe in faulting, tilting of strata, mass extinction, and abrupt change of inferred environment. Sir Charles Lyell and the nineteenth-century uniformitarians argued that missing data of a woefully imperfect geological record would supply the gradual links between events not recorded by rocks.

Moreover, Hutton's commitment to deep time was not unique. As Roy Porter shows in his excellent introduction to William Hobbs's *The Earth Generated and Anatomized* (1715), the potential eternity of the world was a major issue in the great debate on earth systems that peaked in Britain about one hundred years before Hutton wrote his treatise. Hobbs was such an obscure amateur that Porter, after painstaking research, cannot even be sure that the gentleman dismissed for dishonesty as an excise officer, and later cited in tavern brawls and domestic disturbances, is *his* William Hobbs. Hobbs's previously unpublished treatise rests upon the idea, already archaic in his own time as the mechanical world view spread, that the earth is an active organism. Hobbs wrote his work primarily to argue that tides are not caused by the moon's gravitational pull, but by the earth's pulsating heart. In the usual Manichaean division, Hobbs should be among the anti-empirical system builders whom Hutton replaced.

Yet Hobbs, and many of the system builders, supported several of the ideas that Hutton supposedly discovered in the rocks one hundred years later. Hobbs denied that sediments were products of deposition in Noah's flood; he shared Hutton's belief in restorative forces, though he attributed uplift to the same pulsating heart that produced tides; he repudiated Moses and proclaimed a much older earth. He railed against armchair speculation ("verbal science" as he called it) and made many astute observations on rocks and tides. He also eschewed teleology and final causes and was, to that extent, more "modern" than Hutton (in the invalid assessment that assigns worth by anachronistic convergence upon current views). Hobbs had no influence upon science, but his work possesses enormous value both for its own charm and for its illustration of how an untutored man, working far from the literati of London, integrated his own thoughts with common perceptions of his time.

Hutton, like Hobbs, was a system builder. He cited field evidence as an important source of support, but he was no modern empiricist. I do not say this to detract from Hutton's reputation, for there has never been a finer intellect in our profession. His treatise is largely a brilliant methodological argument for inferring unobservable processes from their results preserved in rocks. As such, it is a true landmark and beginning—for past processes are unobservable in principle, and reconstructing the past is geology's primary goal. But we do a disservice to Hutton and all great scientists of the past when we judge them by modern standards and define good science as unfettered observation and logical inference made by minds uncluttered with prejudices of former ages.

Creative science is always a mixture of facts and ideas. Great thinkers are not those who can free their minds from cultural baggage and think or observe objectively (for such a thing is impossible), but people who use their milieu creatively rather than as a constraint—as Darwin did in translating Adam Smith's economics into nature as the principle of natural selection, and as Hutton did in using the principle of final causes to construct a cyclical view of the world.

Such a conception of science not only validates the study of history and the role of intellect—both subtly downgraded if objective observation is the source of all good science. It also puts science into culture and subverts the argument—advanced by creationists and other modern Yahoos, but sometimes unconsciously abetted by scientists—that science seeks to impose a new moral order from without.

III

Biological Determinism

7

Genes on the Brain

Immodest proclamation justly accompanies great discovery; who would gainsay Archimedes shouting "Eureka" through the streets of Syracuse, or announcing that his lever would move the earth if only he could find a place to stand. More often than not, however, immodest proclamation is a cover-up, conscious or not, for failure. When conscious, the tactic can be stunning in its audacity. When unconscious, it is hollow.

Both titles of Lumsden and Wilson's book—and its content—record unconscious failure. They have discovered, they claim, the Promethean fire of our evolution, the key to an understanding of both the origin and the subsequent history of the human mind. This key, they proclaim, is a "largely unknown evolutionary process we have called gene-culture coevolution: it is a complicated, fascinating interaction in which culture is generated and shaped by biological imperatives while biological traits are simultaneously altered by genetic evolution in response to cultural innovation."

In responding to criticisms that human sociobiology, in its debut as the last chapter of Wilson's *Sociobiology* (1975), ignored culture for a crude form of genetic determinism, Lumsden and Wilson have now discovered culture and use it as half

A review of *Promethean Fire: Reflections on the Origin of Mind* by Charles L. Lumsden and Edward O. Wilson.

of a positive feedback loop to explain, with genetics as the other portion, all the essentials of our mental evolution. Lumsden and Wilson summarize their concept of gene-culture coevolution in the following way:

> The main postulate is that certain unique and remarkable properties of the human mind result in a tight linkage between genetic evolution and cultural history. The human genes affect the way that the mind is formed—which stimuli are perceived and which missed, how information is processed, the kinds of memories most easily recalled, the emotions they are most likely to evoke, and so forth. The processes that create such effects are called the epigenetic rules. The rules are rooted in the particularities of human biology, and they influence the way culture is formed. . . .
>
> This translation from mind to culture is half of gene-culture coevolution. The other half is the effect that culture has on the underlying genes. Certain epigenetic rules—that is, certain ways in which the mind develops or is most likely to develop—cause individuals to adopt cultural choices that enable them to survive and reproduce more successfully. Over many generations these rules, and also the genes prescribing them, tend to increase in the population. Hence, culture affects genetic evolution, just as the genes affect cultural evolution.

Promethean Fire is, essentially, a long argument that this unexceptional, and scarcely new, style of evolution can explain what may be the three most important aspects of our own history and current status.

1. Gene-culture coevolution was the trigger for the historical origin of mind in human evolution. It propelled the evolution of increased brain size at a rate perhaps never exceeded for major events in the history of life.
2. Many important universal aspects of human behavior have a genetic basis and set the epigenetic rules that constrain culture.

3. Differences among human cultures, though recent in origin and often deemed superficial, are not free of genetic influence and are usually shaped, or at least strongly influenced, by the efficient process of gene-culture coevolution.

Unfortunately for the high hopes held by Lumsden and Wilson, the first point, while undoubtedly just, is scarcely original and has formed the core of speculations about the evolutionary origin of mind ever since Darwin; the second point, also uncontroversial, is trivial, at least for the examples now available; while the third, controversial and even revolutionary if it could be established, is almost surely false as a general, or even as a common phenomenon.

THE EVOLUTIONARY ORIGIN OF MIND

Lumsden and Wilson begin their book by staking a claim for discovering the origin of mind:

> What was the origin of mind, the essence of humankind? We will suggest that a very special form of evolution, the melding of genetic change with cultural history, both created the mind and drove the growth of the brain and the human intellect forward at a rate perhaps unprecedented for any organ in the history of life. . . .
>
> For the first time we also link research on gene-culture coevolution to other, primarily anatomical studies of human evolution, and use the combined information to reconstruct the actual steps of mental evolution.

The evolution of the human brain must have followed a peculiar pattern strongly implicating some phenomenon like gene-culture coevolution. When we first encounter our ancestors, the australopithecines, in Africa some three million to four million years ago, they had already undergone a major anatomical transformation to upright posture without a concomitant change in their brains, which remained at an ape's characteristic size. Why did these two essential features of our evolution—upright walk and large brains—evolve in such a detached manner, and in this particular sequence? Why did the brain evolve later, after so much of essential human anatomy was already in place?

We have had empirical knowledge of this pattern since the 1920s, when australopithecines were first discovered in South Africa. But the theme of upright walk first, brains second had been correctly surmised, in a speculative way, by many thinkers about human evolution, in part by Darwin himself, but particularly—and with remarkable perspicacity—by his German champion Ernst Haeckel.

Lumsden and Wilson, disregarding this history, stake their own claim for discovery: our brains enlarged and our minds took off only when we entered the positive feedback loop of their newly discovered process: gene-culture coevolution. The speed of our brain's increase then records the accelerative power of positive feedback.

I don't doubt that something like gene-culture coevolution was involved in the evolution of our brain. But then Darwin and Haeckel, and all other major thinkers about human evolution, have made the same argument. In fact, I don't know that any serious theory other than gene-culture coevolution has ever been proposed to explain the sequence of upright posture first, brains later and quickly. The standard account argues that upright posture freed the hands for development of tools and weapons. This evolving culture of artifacts and their attendant institutions of hunting, food gathering, or whatever, then fed back upon our biological (genetic) evolution by setting selection pressures for an enlarged brain capable of advancing culture still further—in short, gene-culture coevolution.

Darwin wrote in *The Descent of Man* (1871):

> If some one man in a tribe, more sagacious than the others, invented a new snare or weapon, or other means of attack or defense, the plainest self-interest, without the assistance of much reasoning power, would prompt the other members to imitate him; and all would thus profit. . . . If the new invention were an important one, the tribe would increase in number, spread and supplant

> other tribes. In a tribe thus rendered more numerous there would always be a rather better chance of the birth of other superior and inventive members. If such men left children to inherit their mental superiority, the chance of the birth of still more ingenious members would be somewhat better, and in a very small tribe, decidedly better.

Ironically, for the man's work is anathema to Wilson, who senses the evil influence of Marxism behind all radical criticism of his sociobiology, the best nineteenth-century case for gene-culture coevolution was made by Friedrich Engels in his remarkable essay of 1876 (posthumously published in the *Dialectics of Nature*), "The part played by labor in the transition from ape to man."

Engels, following Haeckel's outline as his guide, argues that upright posture must precede the brain's enlargement because major mental improvement requires an impetus provided by evolving culture. Thus, freeing the hands for inventing tools ("labor" in Engels's committed terminology) came first, then selective pressures for articulate speech, since, with tools, "men in the making arrived at the point where they had something to say to one another," and finally sufficient impetus for a notable (and genetically based) enlargement of the brain:

> First labor, after it, and then with it, articulate speech—these were the two most essential stimuli under the influence of which the brain of the ape gradually changed into that of man.

An enlarging brain (biology, or genes in later parlance) then fed back upon tools and language (culture), improving them in turn and setting the basis for further growth of the brain—the positive feedback loop of gene-culture coevolution:

> The reaction on labor and speech of the development of the brain and its attendant senses, of the increasing clarity of consciousness, power of abstraction and of judg-

> ment, gave an ever-renewed impulse to the further development of both labor and speech.

Those ignorant of history do, after all, repeat it—especially when there is virtually no other path to follow.

GENETIC UNIVERSALS

In Lumsden and Wilson's version of gene-culture coevolution, genetic predispositions common to all normal humans act in the positive feedback loop by setting epigenetic rules—or biases in learning—that constrain and channel culture. To choose their favorite example, avoidance of incest is a biological imperative of great importance, since the frequency of birth defects rises sharply with closeness of relationship between marriage partners (and reaches a maximum for unions between siblings). Thus, any mechanism discouraging incest would be strongly favored by natural selection and should increase within populations.

Of course, genes are not conscious agents and cannot "tell" their bearers, "Don't copulate with close relatives or you're in for trouble." But "epigenetic" or learning rules with a genetic base might be selected for such an effect. We might, for example, be predisposed by our biology not to develop sexual feelings toward those individuals reared with us in early childhood—familiarity breeds contempt. Since, in ancestral societies with limited mobility and tight kinship bonds, close proximity usually meant close relationship, the epigenetic rule produced its desired biological result. Of course, we could "fool" this rule today by separating siblings or, as some societies do (thus providing the shaky basis in evidence for the form of the rule itself) by raising nonrelatives together and gauging their later sexual aversion.

I am supposed to be a "nurturist" in the great "nature-nurture" debate, but I find nothing upsetting in this notion of biological influence upon human behavior. I suppose I must also emphasize once again, and for the umpteenth time as we all do, that the categories are absurd and that there is no "nature-nurture" debate as such, the pleasant alliteration of the phrase notwithstanding. Every scientist, indeed every intelligent person, knows that human social behavior is a complex and indivisible mix of biological and social influences. The issue is not *whether* nature or nurture determines human behavior, for these factors are truly inextricable, but the degree, intensity, and nature of the constraint exerted by biology upon the possible forms of social organization.

No one doubts that biological universals exist. We must sleep, eat, and grow older, and we are not about to give up procreation; almost all our social institutions are influenced by these imperatives. Therefore, the simple listing of imperatives by Lumsden and Wilson, and the specification of the epigenetic rules they establish, is no defense for a "naturist" bias and no vindication of sociobiology. We must ask instead how shaping and constraining are the universals that can be specified? The answer, at least from the list provided by Lumsden and Wilson, is not very much at all. I therefore find this particular invocation of genetics as a determinant of social behavior both trivial and uncontroversial.

Consider Lumsden and Wilson's entire list of seven items: avoidance of brother-sister incest; learning of color vocabularies; preference of infants for objects of particular shapes and arrangements corresponding to the abstract form of a human face versus various scrambled patterns; the universality of certain facial expressions; preference of newborns for sugar over plain water and for sugars in descending order of sucrose, fructose, lactose, and glucose; anxiety of very young children in the presence of strangers; and phobias, particularly those that respond to ancient dangers (like snakes, running water, and thunderstorms) that no longer threaten us in our modern world.

Item two on color provides a good illustration of why I maintain that these genetic universals offer no threat to the attitude often and mistakenly called the "nurturist" position—that human biology is rarely sufficiently constraining to determine human culture directly and that biology usually permits a wide and flexible range of different cultural possibilities. (The two positions should be called biological determinism and biological potentiality, not naturist and nurturist. We might instead refer to determinists and potentialists.)

A series of fascinating studies has shown that, although light comes to us in a continuously varying spectrum of wavelengths, people in all cultures tend to parse the spectrum into four basic colors: blue, yellow, red, and green. "This beautiful illusion," Lumsden and Wilson write, "is genetically programmed into the visual apparatus and brain." In part, we already know the physiological basis of this bias in learning. Visually active nerve cells in the lateral geniculate body of the thalamus, an important "relay station" between our eyes and the brain's visual cortex, are divided into four types and probably code the light we see according to these four major colors. Now why should any potentialist (or even an old-fashioned, caricatured, exaggerated, nonexistent, tabula rasa nurturist) feel threatened by such a discovery? I find it fascinating. Brain and eye are physical objects with complex properties, not neutral filters. Why, to ensure a potentialist position, should they, by some a priori fiat, have to record color exactly as it comes to us from physics without imposing some constraint evolved from their own biological structure?

In a previous work, *On Human Nature,* Wilson devised an apt metaphor to characterize the issue between determinists and potentialists: the genes hold culture on a leash and debate centers upon the length and tightness of that leash. The question is not merely quantitative, with both sides sharing the same assumptions and differing only in their guesses about one or ten feet in the continuum of leash lengths; for a culture held on a taut one-foot leash (and therefore determined in its manifest properties directly by biology) is a qualitatively and fundamentally different entity from a culture attached by a loose ten-foot cord that cannot specify a particular institution but only a broad range of possibilities. That we see a spectral continuum as four colors, or that babies prefer sugar to water and human faces to scrambled designs, does not dictate cultural patterns with great specificity. The character of the biological universals that we can identify (and we have no reason to think that further research will alter the form of example, though the list will obviously be augmented) suggests that the leash is loose and nonconstraining, though well worth our continued examination. The study of these universals, the one aspect of human sociobiology that has some direct evidence going for it, therefore offers no solace for the determinist bias that forms the soul of sociobiology, and the essence of its claim to be a new and revolutionary science.

DIFFERENCES AMONG HUMAN CULTURES

To establish its importance and to fulfill its reductionist research program, human sociobiology cannot rest upon nonconstraining universals. It must demonstrate that *differences* among cultures, and historical change within cultures—the sources of our most interesting arguments about human nature (are the Chinese really . . . ?)—are also genetically driven. The crux of the determinist–potentialist debate lies squarely here, for genetics (and therefore Darwinian biology strictly conceived) can say little about human cultural diversity and change if it only underlies some universal patterns and exerts little constraint upon the incredible richness of detail that so fascinates us and *is* the subject matter of the social sciences. Lumsden and Wilson know that they must deliver here, or the revolution of human sociobiology dies aborning. They also know that no concrete information supports their hope. Consequently, they devoted their last book *(Genes, Mind, Culture)*, to constructing a mathematical model showing that their favored results could occur in principle under a set of dubious assumptions (*Promethean Fire* is a popularized abstract of this technical and mathematical work). Their models fared poorly under critical review[1] and *Promethean Fire* relies wholly on verbal appeals to plausibility.

The problem faced by Lumsden and Wilson is deep indeed. Adjacent cultures differ sharply in their basic beliefs and institutions; historical change within cultures may drive them from heights of power to depths of impotence on a

[1]For example, J.M. Smith and N. Warren, "Models of Cultural and Genetic Change," in *Evolution*, vol. 36 (1982), p. 620.

time scale of hundreds of years, or even generations (consider the history of Islam from its days of glory late in our first millennium, to its time of despair during the height of Western colonialism, to its current rebirth). How, given this potential (and often realized) speed and depth of change, can we invoke the slow process of Darwinian natural selection as a shaping force?

Lumsden and Wilson's answer, of course, is the accelerative power of positive feedback operating in gene-culture coevolution. With culture as a boost, genetics can play a powerful role in so short a time. But abstract models are one thing; their realization is another matter. All manner of implausibilities can be modeled when assumptions are chosen to guarantee the result. I doubt that biological change can play an important role in cultural diversity when we know that nongenetic forces of climate, conquest, and the invention and spread of new technologies exert so controlling an influence. History rather than genetics must be the ground for our search to understand cultural diversity and change.

Where evidence does exist, Lumsden and Wilson must admit that the data do not suit their hopes. They write, for example:

> Even the caste system of India, which is the most rigid and elaborate on Earth and has persisted for two thousand years, is maintained largely by cultural conventions. So far as is known (although the matter has never been thoroughly studied), members of different castes differ from one another only slightly in blood type and other measurable anatomical and physiological traits.

Lumsden and Wilson end their chapter on rules of mental development with these words:

> The theory of human nature that prevails in the end will be the one that aligns social behavior and history with all that is known about human biology. It will correctly and uniquely characterize the known operations of the human mind and the patterns of cultural diversity. That

is the grail towards which many scholars toil, despite frustration and not infrequent humiliation. At issue are the very limits of the natural sciences. Is the quest a fool's errand?

I find nothing in this somewhat grandiloquent statement to offend any potentialist, though I have personal doubts about "correct and unique" characterization, given the vagaries of history. The quest is no fool's errand and we all seek to gain the final understanding of Parsifal. We simply differ in our views about the relative importance of biology in this future alignment of social behavior, history, and genetics. I suspect that biology will not play an important part in explaining "patterns of cultural diversity"; to this extent, the natural sciences do meet their limit. But why should this result be sad or disappointing? It does not mean that knowledge must remain fragmentary; it simply holds that the correct empirical equation will grant a large coefficient to history and a small part to genetics.

THE ASSUMPTIONS OF SOCIOBIOLOGY

In the absence of evidence for their most important claim, and in the face of some information and several strong prima facie arguments against it, we must ask what infuses Lumsden and Wilson with such confidence about the promise of sociobiology as a key to explaining human cultural diversity. Here we encounter the methodological premises of the sociobiological research program in Wilson's version. Sociobiology has engendered a great deal of debate in several disparate fields since Wilson published his book of the same name in 1975. Political discussion about the uses of biological determinism have often masked the deeper methodological objections that call the whole enterprise into question, whatever its implications. (I have pursued the political debate myself, and certainly do not abjure it, but we must recognize that a more fundamental criticism questions the essential style of sociobiological argument itself as an appropriate application of evolutionary theory.)

To substitute biology for history in the absence of evidence requires an a priori faith that genetic explanations are, in some ultimate sense, preferable. Such a position emerges from the old-fashioned reductionism espoused so strongly by Lumsden and Wilson. A hierarchy of sciences runs from hard to soft, quantitative to qualitative, firm to squishy, from physics through biology to the loose domain of social sciences. We should rejoice any time we can force an explanation down from the realm of a soft science to a harder domain. Genetics really is better than history as a scientific explanation. At times, Lumsden and Wilson's reductionism can become downright militant:

> The bridge between biology and psychology is still something of an article of faith, in the process of being redeemed by neurobiology and the brain sciences. Connections beyond, to the social sciences, are being resisted as resolutely as ever. The newest villain of the piece, the embattled spearhead of the natural-science advance, is sociobiology.

I have attacked this style of reductionism in other articles[2] and will not rehearse the arguments here—except to say that if no important genetic differences underlie cultural diversity and change, then an intrinsic preference for a style of argument must bow to the constraints of information.

In a classic error of reductionism, Lumsden and Wilson write:

> To many of the wisest of contemporary scholars, the mind and culture still seem so elusive as to defeat evolutionary theory and perhaps even to transcend biology. This pessimism is understandable but, we believe, can no longer be justified. The mind and culture are living phenomena like any other, sprung from genetics, and their phylogeny can be traced.

[2]See especially essays 10 and 16.

But historical origin and current function are different aspects of behavior with no necessary connection. Of course the mind is sprung from genetics (or at least involved fundamental genetic changes in the evolution of brain), but such a statement about history does not guarantee a biological basis for current cultural diversity since the springboard, once installed, may set a common genetic basis, while cultural diversity then develops as a historical overlay. I am also amused by the giveaway admission of reductionist bias—the claim that any denial of evolutionary theory as a proper locus for the explanation of culture must denote pessimism. After all, the "wise scholars" of this statement are not intoning *ignorabimus* (we shall never know), but merely claiming that culture will achieve its primary explanation from disciplines other than biology.

Another aspect of reductionism underlies Lumsden and Wilson's application of sociobiology to culture, and helps to explain why so many evolutionary biologists, who have no political feelings about the matter and who (as professionals) share Wilson's hope for the advancing hegemony of biology, reject the specific mode of argument used in this book. Wilson is our leading student of the behavior of social insects, those marvelous creatures of little brain whose behavior can be atomized into a set of individual traits, each treated independently. Ants behave, in many essential respects, as automata, but human beings do not and the same methods of study will not suffice. We cannot usefully reduce the human behavioral repertoire to a series of unitary traits and hope to reconstruct the totality by analyzing the adaptive purpose of each individual item.

First of all, we can't come close to agreement on a proper atomization, probably because behavioral wholes are not simple aggregates of any set of small parts (is "xenophobia" really a "thing;" is it the proper category for analyzing both a young infant's aversion to strangers and an adult's professed hatred of other races?). Second, and perhaps more important, we have no reason to believe that each item maintains its particular form (or set of alternative states) as an adaptation engendered by natural selection. Yet Lumsden and Wilson are committed to a strict version of Darwinism that equates genetic change with adaptation and therefore must analyze all important behavioral differences according to their advantages in the varying environments occupied (or once inhabited when the trait arose) by the cultures under analysis. Thus, for example, religion, despite all its complexities, is labeled "a powerful device by which people are absorbed into a tribe and psychically strengthened."

Lumsden and Wilson have even constructed a terminology for their atomization. They propose to call each unit of human behavior a "culturgen," derived from the Latin for "creating culture," but clearly recalling both elements of the positive feedback loop between gene and culture. Their analysis then traces the spread of each culturgen, considered independently, via the adaptive force of natural selection. Consider their words on the supposedly self-evident nature of culturgens. (Even these examples are dubious reifications, except perhaps for the third, and they are chosen as best illustrative cases. Imagine what happens when we come to more culture-laden concepts like stereotyping and sexual behavior.)

> Many culturgens are naturally distinct and would stand with or without theory. The preference for incest, for example, exists as a clear alternative to the preference for outbreeding. Women tend to carry infants on their left side close to the heart, a practice easily distinguished from other modes of infant transport. To raise the eyebrow in greeting is a gesture distinct from other facial signals.

Lumsden and Wilson then confuse actual history with their preferred methodology by claiming that since genetics began with Mendel's single traits, the study of human behavior must also start with unitary culturgens and build up the entire repertoire sequentially: "By detecting and analyzing large numbers of such single-gene differences, great and small, a picture of the full genetic blueprint can gradually be assembled. This is the way genetics has proceeded from the garden plots of Mendel to the mighty enterprise it is today."

The problem of adaptation extends far beyond this dilemma of atomization because a strong argument can be advanced that most "things" done today by the brain could not have evolved originally as direct adaptations connected with its evolutionary increase in size.

The debate has a long pedigree and goes back to a bitter disagreement between Darwin and the codiscoverer of natural selection, Alfred Russel Wallace. Wallace, a true pan-selectionist in the Wilsonian manner (as Darwin was not), advanced the curious argument that natural selection could account directly for every trait in the evolution of all organisms except for the human brain, which required a divine assist. He was accused of lacking courage, of failing to extend his system to the final and most important step of all. I cannot analyze the psychology of his reluctance; but, ironically, the logic of his argument springs not from his religion or spiritualism, but from his pan-selectionism itself.

Wallace, in an uncommon attitude for nineteenth-century white scientists, was a nonracist who truly believed in the equal mental capacities of all people. But he was a cultural chauvinist who did not doubt the overwhelming superiority of Western European institutions. Now, if natural selection constructs organs for immediate use and if brains of all people are equal, how could natural selection have built the original "savage's" brain (his terminology)? After all, savages have capacities equal to ours, but they do not use these abilities in devising their cultures. Therefore, natural selection, a force that constructs only for immediate utility, cannot have fashioned the human brain.

Darwin was flabbergasted. He wrote to Wallace: "I hope you have not murdered too completely your own and my child." His simple counterargument, born from his pluralistic attitude toward the variable power of natural selection, held (if I may use an anachronistic modern terminology): the brain is a very complex computer. I have no doubt that natural selection produced its increase in size and mental power. Selection probably built our large brain for a complex series of reasons, now imperfectly understood. But whatever the immediate reasons, the enlarged brain could perform (as a consequence of its improved structure) all manner of operations bearing no direct relation to the original impetus for its increase in size. I may put a computer in my factory only to issue paychecks and keep accounts, but the device can (as a consequence of its structure) also calculate pi to 10,000 places and perform a factor analysis on the correlation matrix of human culturgens.

Historical origin and current function are different properties of biological traits. This distinction sets an important general principle in evolutionary theory. Features evolved for one reason can always, by virtue of their structure, perform other functions as well. Sometimes the principle is of minor importance, for the directly selected function may overwhelm any side consequence. But the opposite must be true for the brain. Here, surely, the side consequences must overwhelm the original reasons—for there are so vastly more consequences (surely by orders of magnitude) than original purposes.

Consider only, for example, our knowledge of personal mortality. Nothing that our large brain has allowed us to learn has proved more frightening and weighty in import. Surely no one would argue that our brains increased in order to teach us this unpleasant truth. Yet consider the impact of this knowledge upon a diverse range of human institutions, from religion to kingship and its divine right. The specific forms of religion need not be seen as direct adaptations "for" tribal cohesion; much of culture may arise as responses to the curious and unpredictable side consequences of a large brain. Thus, and particularly for the human brain (that key organ of human sociobiology), the adaptive analysis of culturgens cannot unlock their meaning, since most did not arise for adaptive reasons and are not, therefore, "items" for Darwinian analysis.

The message of this assertion is not, as adaptationists often charge, the pessimistic statement that reasons for concepts and institutions of culture must always remain elusive. We must simply shift our focus from guesses about adaptation to direct study of structure. Several years ago, Francis Crick said to me, after a talk I had given at the Salk Institute, "The trouble with you evolutionary biologists is that you are always asking 'why' before you understand 'how.' " Ten years before, I would have dismissed this comment as the misunderstanding of a reductionist molecular biologist who simply didn't comprehend the distinctive character of evolutionary reasoning. But now I understand and agree with Crick—for we have to grasp the "hows" of structure before we can even ask whether or not a direct "why" exists.

Despite these fatal flaws of atomization and adaptationism, Lumsden and Wilson remain confident of their future as sociobiologists. They write:

> Human sociobiology is in approximately the same position as molecular biology in its earliest days. That is, several key mechanisms have been identified, enough to explain elementary phenomena in a new and more precise way. The subject is still rudimentary, but if both biology and culture are to be taken into account, it seems the only way to go.

But the success of one field after a painfully slow beginning does not guarantee that all disciplines in a similar state have a rosy future; as with species and businesses, most never get very far. I think that Lumsden and Wilson have missed the crucial difference between molecular biology and human sociobiology. For molecular biology, the reductionistic research program really did work, triumphantly (though it has now reached limits in considering the cohesion of entire genomes). After all, molecular biology is, to a large extent, chemistry. But the same reductionistic strategy will not work for human culture. When we talk of falling in love as a form of chemistry, we speak only in metaphor. And that is a profound difference, both for poets and scientists.

8

Jensen's Last Stand

JENSEN THEN AND NOW

There are many styles of retreat in the face of failure. As a first and most forthright strategy, one can simply be humble and contrite. Clarence Darrow once stated that if God really existed after all, and if he were, following his death, arraigned before God as judge with the twelve apostles in the jury box, he would simply step up to the bench, bow low, and say: "Gentlemen, I was wrong." In a second, intermediate strategy—the stiff upper lip—one looks upon the bright side (or sliver) of admitted adversity. When Robert FitzRoy, Darwin's captain on the *Beagle,* learned that Jemmy Button, the Fuegian native he had trained in English ways, had "reverted" completely to old habits within months of his return, FitzRoy took refuge in the thought that "a shipwrecked seaman may hereafter receive help and kind treatment from Jemmy Button's children; prompted, as they can hardly fail to be, by the traditions they will have heard of men of other lands." As a third tactic, one proclaims triumph and punts hard. I remember Senator Aiken's brilliant solution to the morass of Vietnam—that we should simply declare victory and get out.

Arthur Jensen has published an 800-page manifesto embodying this third strategy. To understand why it represents a retreat—and a failed retreat at that—we must review the

A review of *Bias in Mental Testing* by Arthur R. Jensen.

history of its genesis. In his notorious article of 1969,[1] the founding document of "Jensenism" as a public issue, Jensen maintained that compensatory education must fail because the black children that it attempted to aid were, on average, genetically inferior to white children in intelligence. He based his claim on a strong form of genetic argument: the heritability of IQ, he maintained, had been adequately estimated at about 0.8 among whites; therefore, the 15 point average difference in IQ scores between blacks and whites must be largely innate in origin.

The intervening decade between this article and the present book has not been kind to Arthur Jensen. First of all, the estimate of heritability, depending heavily on Sir Cyril Burt's faked data,[2] is clearly too high. Second, and more important, the value of heritability *within* either the white or the black population carries no implication whatever about the causes for different average values of IQ *between* the two populations. (A group of very short people may have heritabilities for height well above 0.9, but still owe their relative stature entirely to poor nutrition.) Geneticists treat variation within and between groups as entirely different phenomena; this is a lesson taught early in any basic course. Jensen's conflation of these two concepts marked his fundamental error.

I assume that Jensen now understands where he went seriously astray. The present book bypasses the issue of heritability entirely. In dismissing this previous bulwark of his system in just two paragraphs, Jensen simply states that the matter is too complicated for treatment here, though he treats in intricate detail some of the most arcane and complex issues of psychometrics throughout his 800 pages. Heritability, he argues, "is a highly technical and complex affair involving the principles and methods of quantitative genetics." "Because even an elementary explication of heritability analysis is be-

[1]A.R. Jensen, "How much can we boost IQ and scholastic achievement?" *Harvard Educational Review* (1969) 39:1–123.

[2]L.S. Hearnshaw, *Cyril Burt, Psychologist* (Ithaca: Cornell University Press, 1979).

yond the scope of this book, the interested reader must be referred elsewhere." His list of appended references includes not a single one of several cogent critiques directed against his original thesis.[3]

Moreover, Jensen now claims that we have no need to talk about genetics anyway: "Because we have no estimate of the individual's genotype that is independent of the test score, there is really no point in estimating genotypic values." In fact, he now virtually argues that the subject of causation should be dropped entirely: "The constructors, publishers, and users of tests are under no obligation to explain the causes of the statistical differences in test scores between various populations. They can remain agnostic on that issue." Yet his acceptance of this very obligation is the motivating theme of his 1969 article.

Am I not being unkind in bringing all this up? A man should be allowed to change his mind with grace, and to save face in his expiation. But Arthur Jensen hasn't altered his basic tune at all. He is simply using a different, and more indirect, argument to prop up the same old claim. And he has buried the central fallacies of that argument so deeply among the apparent rigor of these 800 pages of lists, figures, and charts that no commentator in the mass media managed to ferret them out.[4] Jensen's fundamental claim is still about innateness. Indeed, it is still the same claim: blacks are less intelligent than whites and this difference cannot be attributed to environment.

In reasserting his 1969 claim in its more indirect form, Jensen constructs an argument of three parts:

1. The average difference in IQ between whites and blacks is about 15 points, or one standard deviation. Other tests of intelligence show comparable differences.

[3]L.J. Kamin, *The Science and Politics of IQ* (New York: John Wiley, 1974); see also articles by R.C. Lewontin, J. Hirsch, and D. Layzer in N.J. Block and G. Dworkin (eds.), *The IQ Controversy* (New York: Pantheon, 1976).

[4]*Time*, September 24, 1979, p. 49, and *Newsweek*, January 14, 1980, p. 59. I also read several technical reviews after this article appeared, and none of them found the central fallacy either.

2. The tests are unbiased.
3. IQ (and other valid tests of the same mental attributes) measure something that we can legitimately call "intelligence."

Note that although the argument says nothing about genetics or innateness, it seems to lead inexorably in that direction. After all, if blacks perform more poorly than whites on unbiased tests that measure intelligence, then blacks must be less intelligent than whites for reasons unrelated to environmental deprivation (our usual supposition for the cause of "bias," as we use the term in the vernacular). What reasons besides innateness are left?

Of the three arguments, only the first, an undisputed fact, compels assent. It also leads nowhere because it implies nothing whatever about the reasons for the difference. The tests themselves may record nothing of interest; potential reasons for difference span the entire range from pure environmental imposition to pure innateness. Jensen's argument becomes meaningful and controversial only if the second and third points are valid in our vernacular understanding. I believe that they are not valid and that this book, despite its wealth of interesting technical detail for psychometricians, therefore contains no important or general message to enlighten the general concern that we all must feel for the issue of human potential.

THE MEANINGS OF BIAS

Jensen has titled his book *Bias in Mental Testing,* and most of its ample length is dedicated to proving either that bias doesn't exist, or that bias can be recognized and corrected if it does exist. And I'm sure that he's right.

The last paragraph may, at first glance, seem utterly destructive to my case, but it isn't. Things are seldom what they seem in statistics, and a layman's understanding of the field has been plagued by important differences between vernacular and technical meanings of terms. "Significance" and "discrimination" may provide the two most notable cases of difference between English vernacular and statistician's jargon, but "bias" belongs in the same category. In proving that tests are not biased, Jensen speaks to statisticians' interests and not at all to what the public understands by the common charge that IQ tests are biased.

Average black IQ in America is about 85, average white IQ about 100. The charge of bias, in our ordinary understanding of the word, holds that this poorer performance of blacks is a result of environmental deprivation relative to whites, and does not reflect inherent ability. The vernacular charge of bias (I shall call it V-bias) is linked to the idea of fairness and maintains that blacks have received a poor shake for reasons of education and upbringing, rather than of nature.

For two months, I have tabulated every use of the word I have seen in the popular press, and all have conformed to this understanding. *The New York Times,* for example, reported that Federal Judge Robert Carter "has ruled that the examination [for police officers] was biased" because so few black and Hispanic applicants scored among the highest grades. And *The Sacramento Bee* outlined Judge Robert F. Peckham's decision to ban IQ tests as a criterion for placing children in EMR (educable mentally retarded) classes in California. Peckham was disturbed by the preponderance of blacks in such classes and ruled "that there probably could not be any substantial disproportion of blacks . . . if the process of selection was unbiased." In other words, both assumed that the rarity of high scores for blacks or low scores for whites does not reflect natural aptitude fairly. If Jensen had proved that tests are unbiased in this sense (V-bias), he would have made an important and deeply troubling point.

But "bias," to a psychometrician, has an utterly different and much narrower meaning—and Jensen addresses himself only to this technical sense (which I shall call S, or statistical bias). An intelligence test is S-biased in assessing two groups only in the following circumstance: Suppose that we plot the scores on an intelligence test, showing them in relation to what we wish to predict from the test—job performance or school grades, for example. The test is unbiased in a statistician's sense if and only if points for blacks and whites fall along the

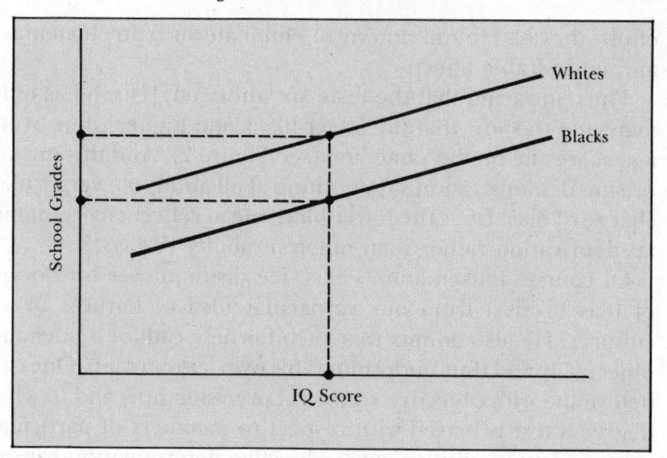

Figure 1. A test with S-bias. The same IQ score predicts different grades for blacks and whites.

same line—that is, if lines for blacks and whites, plotted separately, do not differ in slope, y-intercept (the point of intersection between the lines and the vertical axis) or standard error. If this seems confusing, consider Figure 1, an example of intercept bias. Whites and blacks have the same slope, but whites have a higher y-intercept.

It is not difficult to see why psychometricians want to rid themselves of S-bias; for in an S-biased test, the same score yields different predictions based upon group membership. In Figure 1, an IQ of 100 predicts poorer grades for a black than for a white. No sensible tester wants to construct an instrument in which the same score means different things for different kinds of people.

Jensen devotes most of his book to showing that S-bias does not affect mental tests (or that S-bias can be corrected when it does exist). Yet I found nothing surprising in his densely documented demonstration that tests are unbiased in this sense. It would be a poor reflection indeed on the technical competence of psychometricians if, after nearly a century of effort, they had found no way to eliminate such an elementary and undesirable effect.

Thus, in saying that the tests are unbiased, Jensen has only managed to show that the lower black and higher white average scores lie on the same line (see Figure 2). And this unsurprising demonstration says nothing at all about the vernacular charge of bias: Does the lower black mean reflect environmental deprivation rather than inherent ability (V-bias)?

Of course, Jensen admits this. He distinguishes his notion of bias (S-bias) from our vernacular idea of fairness to all cultures. He also admits that such fairness cannot be defined objectively and thus undermines his own larger case: "One can determine with objective statistical precision how and to what degree a test is biased with respect to members of particular subpopulations. But no such objective determination can be made of the degree of culture-loadedness of a test. That attribute remains a subjective and, hence, fallible judgment. . . . The term 'bias' is to be kept distinct from the concept of fairness-unfairness."

Yet these brave words are obfuscated or diluted throughout the book for three reasons. First, although he makes the distinctions fairly and forthrightly, he buries them on two pages in the middle of a lengthy work, and does not emphasize them thereafter. Second, Jensen correctly points out that some kinds of S-bias may have an environmental source. (The higher y-intercept of whites in Figure 1, leading to higher school grades among whites than blacks for the same test score, may reflect environmental advantages not measured by the test.) Thus, the concepts of S-bias and environmental difference become subtly conflated (and one easily slips into the false conclusion that tests without S-bias cannot record differences of environment)—even though the existence of S-bias is irrelevant to the key question about environment that has sparked the whole debate: does the lower black mean reflect environmental disadvantages? Indeed, Jensen's 1969 article argued that the mean differences could not be attributed to V-bias because they are primarily genetic in origin.

Third, and most importantly (and annoyingly), Jensen, after making clear distinctions between S-bias and "culture fair-

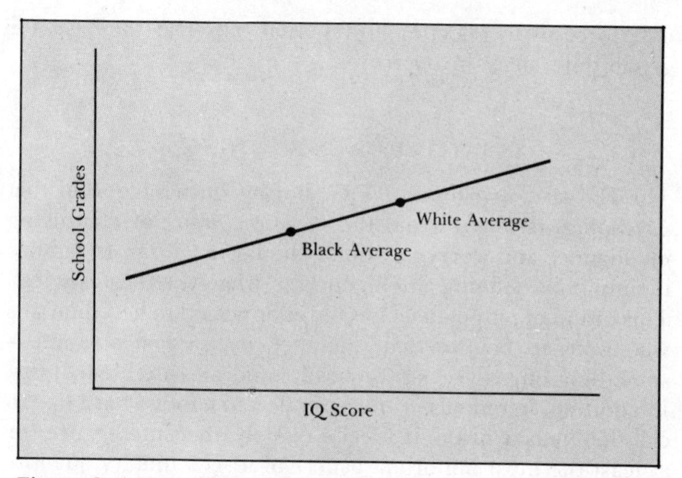

Figure 2. A test without S-bias.

ness," then proceeds to confuse the issue completely by using "bias" in its ordinary vernacular sense over and over again. He speaks, for example, of "the hypothesis that, when the Stanford-Binet is administered to any population other than the original normative sample, the different population should score lower than the normative sample because of cultural biases." In another place he speaks of two tests that were "culturally biased" to award rural or urban children the higher score. These passages speak of mean differences between groups that may lie on the same line in plots of test scores versus criterion. The potential cultural bias is therefore V-bias. But Jensen has told us that we may not use the term "bias" for such a concept.

In short, the primary content of this book is simply irrelevant to the question that has sparked the IQ debate and that Jensen himself treated in his 1969 article: what does the lower average score of blacks mean? His concept of bias (S-bias) does not address this issue. Yet, since this issue is intimately associated with our vernacular meaning of bias, nonstatistical reviewers (in *Time* and *Newsweek,* for example) have been consistently confused into believing that Jensen's voluminous data force us to reject environmental causes as the basis for group differences in IQ scores.

IS INTELLIGENCE A "THING"?

The Harvard psychologist E.G. Boring once suggested that psychologists might avoid the vexatious issue of identifying intelligence and worrying about whether or not tests capture it simply by defining intelligence as whatever the tests test. This brand of pragmatism has never appealed to hereditarians who want to believe that intelligence is a real attribute—something objective, "in the head," and measurable by tests. In addition, hereditarians have tended to adopt what I like to call the "fallacy of the ladder"—namely, that intelligence (or at least the most important aspect of it) is a unitary quantity that can be assigned as a single number to each individual. People may then be ranked on an ascending ladder from ape to Einstein—a single scale that captures the most important aspect of their potential. The entire concept of IQ is rooted in this fallacy.

The only important theoretical justification that psychometricians have ever offered for viewing intelligence as a real attribute that can be measured by a single number (IQ) arises from an arcane subject called "factor analysis." Jensen argues correctly that factor analysis is "absolutely central to any theoretical understanding of intelligence." Factor analysis in psychometrics has received virtually no discussion outside professional circles. This is particularly unfortunate since the history and theory of IQ testing cannot be understood without reference to it.[5]

[5]My book, *The Mismeasure of Man* (New York: W.W. Norton, 1981) presents the first popular explanation of factor analysis in relation to the history of mental testing. In my entire career, I have never had such difficulty rendering a technical subject in layman's terms. I succeeded (if I did) only because I found a geometric (therefore pictorial) way to display the formulations of matrix algebra. This discussion follows the same strategy and was the prototype for my later elaboration.

In arguing that intelligence is a single, definable "thing," and that blacks possess less of it than whites, Jensen has resurrected the original form of Charles Spearman's argument for factor analysis, a hypothesis that had been (or so I had thought) at best moribund since L.L. Thurstone made his devastating criticisms during the 1930s.

Charles Spearman developed factor analysis in 1904 to deal with an interesting, though unsurprising, observation: if several mental tests are given to a group of people, scores tend to be positively correlated—that is, people who do well on one kind of test tend to do well on others. Spearman wondered whether there might not be some common factor underlying this tendency for similarity of performance in each individual. He compiled what statisticians call "a matrix of correlation coefficients" and extracted from it a single number, which he called *g* or general intelligence. Spearman concluded:[6]

> All examination in the different sensory, school, and other specific faculties may be considered as so many independently obtained estimates of the one great common Intellective Function.

A correlation coefficient is a measure of association between two tests given to several people. Its value ranges from -1.0 (a good score on one test implies an equally bad score on the other) through 0.0 (score on one test allows no prediction about score on the other) to 1.0 (good score on one implies equally good score on the other). Most correlation coefficients between mental tests are positive, but nowhere near a perfect value of 1.0—that is, people who do well on one test *tend* to do well on the other, but one score doesn't predict the other perfectly and not everyone will do well on both tests. A matrix of correlation coefficients is simply a table that lists all the individual coefficients for a set of several tests.

Spearman believed that his *g* represented a physical entity.

[6]C. Spearman, "General intelligence objectively defined and measured," *American Journal of Psychology* (1904), 15:201–93.

He identified it with a general inborn "cerebral energy," and argued that blacks and poor people had less of this energy than whites and upper-class people. Sir Cyril Burt, Spearman's successor as professor of psychology at University College, London, devoted the largest part of his career to defending the hereditary interpretation of *g*. We now know he used partly faked data to do so. But is *g* really an entity at all? To appreciate why it is not, we must understand how *g* is extracted from a set of correlation coefficients. Fortunately, this can be explained in a simple, geometrical way.

A set of correlation coefficients may be represented as a group of vectors (lines) radiating from a common point. With some (conceptually unimportant) simplification, we draw these lines as equal in length and state that the correlation coefficient between any two tests is given by the cosine of the angle separating the two vectors for these tests. (This matches our intuitions well. Two perfectly correlated tests have overlapping vectors—the cosine of zero degrees is 1.0. Two independent tests have vectors at right angles—the cosine of 90 degrees is zero. The closer any two lines, the higher their correlation coefficient.) In Figure 3, I consider four tests (two verbal and two arithmetic) in two-dimensional space. All four tests are positively correlated (any two vectors are separated by an angle less than 90 degrees), but verbal and arithmetic tests form separate subclusters (the two verbal tests are more strongly correlated with each other than either is to any arithmetic test). This represents the usual situation of positive correlation among all tests with a tendency for subclustering among tests of common character.

Now Spearman's *g* is simply what statisticians call the "first principal component"[7] of this set of vectors. It is the line of

[7]Technically, a first principal component is not the same thing as the first axis of a factor analysis done in the principal components orientation. In my diagram, I work in two dimensions, fit two axes and resolve all the information. This is called principal components analysis. In true factor analysis, one decides beforehand to abandon some information and to work in a space of reduced dimensionality. But the first principal component and the first factor axis in principal components orientation play the same conceptual role and

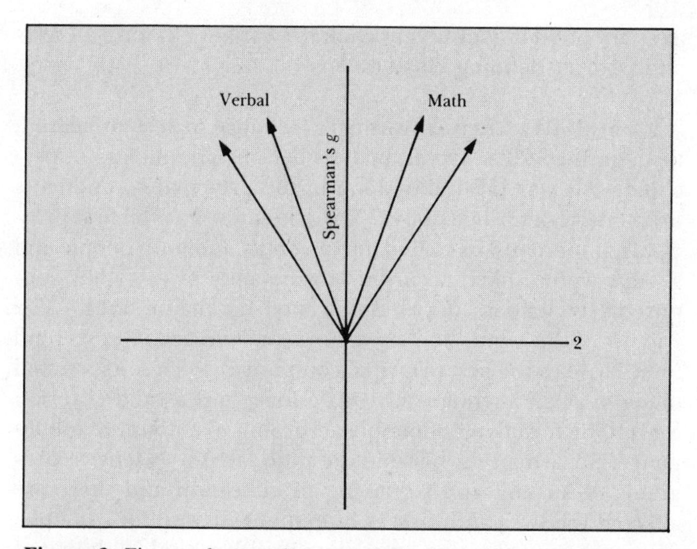

Figure 3. First and second axes, principal components orientation.

best fit, the single axis that "explains" more information in all the vectors than any other line in any other position could. In our figure, it runs (unsurprisingly) right through the middle of the cluster. Since vectors for all tests tended to lie near this first principal component, Spearman thought he had discovered an underlying, common intelligence that each test measured only imperfectly. Hence he called this axis *g,* or general intelligence. In our two-dimensional plots of these four vectors, we may fit a second principal component at right angles to the first. The projection of vectors upon the second component does identify a slight separation of the verbal from the arithmetic cluster, but the effect is small because the vectors are primarily resolved into the first component *(g).* In real data, the effect is often completely erased by patterns of variation and errors of measurement. Hence, principal component

---

differ only in mode of calculation: they are "best fit" axes that resolve more information in a set of vectors than any other axis could.

axes are good at identifying common variance among all tests and poor at defining clusters.

From 1904, when Spearman's seminal article appeared, through the 1930s, factor analysis became an industry in psychology; it was invariably done in the principal components orientation described above. The *g* identified as the first principal component was reified into an entity and both people and groups were ranked according to the amount of *g* they supposedly contained. (Cyril Burt called his major book "The Factors of the Mind.") Since vectors for standard IQ tests tend to lie close to the first principal component, such tests seemed to serve as an adequate surrogate for *g* and a valid criterion for unilinear ranking of people according to amount of intelligence. Spearman's *g* became the rationale for extensive programs of sorting and streaming in education and therefore affected the lives and careers of millions. It was, for example, the primary justification for sorting British schoolchildren into separate schools at age eleven (when *g* had stabilized and before "group factors" of specialized abilities became important). And, to use a faddish phrase of the moment, the "bottom line" of that sorting, no matter what its official rationale, was "smart" (20 percent) and "dumb" (80 percent).

In the 1930s, the American statistician and psychologist L.L. Thurstone virtually destroyed this edifice with a simple and elegant argument. He pointed out that the principal components orientation for axes had no theoretical, mathematical, or psychological necessity. It merely represented one of a literally infinite number of possible positions for the placement of axes through a swarm of vectors. Where you place the axes depends upon what you want to learn. Given our deep and subtle prejudices for unilinear ranking and notions of progress, and our not so subtle preferences for ordering people by inferred "value" (with one's own group invariably most worthy), it is not surprising that principal components seemed the most "natural," indeed the only proper way to perform factor analysis. But, Thurstone argued, suppose we are most interested in locating clusters of more specialized abilities, not in finding some inchoate, common variance. Then it would be better to

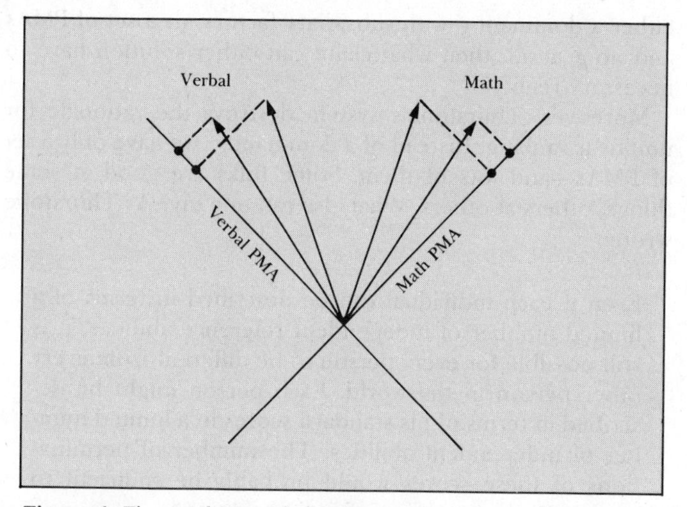

Figure 4. First and second simple structure axes; *g* has disappeared.

place axes near the clusters themselves, in an orientation that Thurstone called "simple structure."

Figure 4 shows a simple structure solution for the same four vectors illustrated previously. Note that verbal and arithmetic clusters are now clearly separated by high projections on one axis with correspondingly low projections on the other. Thurstone used these simple structure axes to identify what he called "primary mental abilities," or PMAs. (He too committed the fallacy of reification and called his simple structure axes—and his major book—"Vectors of Mind.")

Simple structure axes are every bit as good, mathematically, as principal components. They resolve the same amount of information and can be massaged to yield equally cogent psychological interpretations. But note what has happened to *g*, the supposedly ineluctable and innate quantity of general intelligence. It has disappeared; it just isn't there any more. Instead of a pervading and dominating general intelligence and some secondary factors, we now have a set of PMAs. The data have not changed one whit. If the same data can yield either a dominant *g* with subsidiary factors, or a set of PMAs and no *g* at all, then what claim can either solution have to necessary reality?

Moreover, Thurstone's system destroys the rationale for unilinear ranking. Instead of a dominant *g*, we have only a set of PMAs—and lots of them. Some folks are good at some things, others at others. What else can one say? As Thurstone wrote:

> Even if each individual can be described in terms of a limited number of independent reference abilities, it is still possible for every person to be different from every other person in the world. Each person might be described in terms of his standard scores in a limited number of independent abilities. The number of permutations of these scores would probably be sufficient to guarantee the retention of individualities.[8]

In defending his third essential claim—that IQ tests measure something we may legitimately call intelligence—Jensen has resurrected the original form of the Spearman-Burt argument for *g* and the principal components solution. His commitment to the idea of intelligence as a single quantity, distributed in varying amounts among God's creatures, is manifest in the most naïve bit of writing about evolution I have seen in years. Jensen would actually extend *g* throughout the animal kingdom, resurrecting for it the evolutionary ladder that Lamarck advocated, but that Darwin pulled down in showing us that phylogeny is a copiously branched tree, not a ladder of progress. Jensen writes:

> The common features of experimental tests developed by comparative psychologists that most clearly distinguish, say, chickens from dogs, dogs from monkeys, and monkeys from chimpanzees suggests that they are roughly scalable along a *g* dimension . . . *g* can be viewed

[8]L.L. Thurstone, *The Vectors of Mind* (University of Chicago Press, 1935).

as an interspecies concept with a broad biological and evolutionary base culminating in the primates.

But primates are no culmination of anything, just a limb on the mammalian tree; and chicken-dog-monkey-chimp is simply not an evolutionary sequence. Jensen's earlier statement about "different levels of the phyletic scale—that is, earthworms, crabs, fishes, turtles, pigeons, rats, and monkeys" is even more risible, especially when we recognize that modern bony fishes evolved more than 100 million years after turtles and that the evolutionary connection of crabs and vertebrates is not from one to the other, but through some unknown common ancestor that lived more than 600 million years ago. "*The* turtle," whatever that means since they come in hundreds of species, is not, as Jensen claims, "phylogenetically higher than the fish" (meaning even less since rivers, lakes, and oceans contain some 20,000 species of fishes).

Jensen is not even content with *g* as a criterion of ranking in this world. He would extend it throughout the universe! "The ubiquity of the concept of intelligence is clearly seen in discussions of the most culturally different beings one could well imagine—extraterrestrial life in the universe. . . . Can one easily imagine 'intelligent' beings for whom there is no *g*, or whose *g* is qualitatively rather than quantitatively different from *g* as we know it?" With such devotion to quantified, unilinear ranking of intelligence through the universe, it is not surprising that Jensen, in analyzing mere mortals of a single species, would resurrect the Spearman-Burt argument and elevate it from one model among many to necessary reality.

Of course, Jensen is not unaware of Thurstone's critique, and he does discuss it at length. His defense of *g* as dominant and ineluctable arises from a general observation about Thurstone's rotated simple structure axes. In Figure 4, note that the simple structure axes do not lie within the clusters, but outside them. This occurs because the clusters themselves are positively correlated (separated by an angle of less than 90 degrees) while the simple structure axes are defined as mathematically uncorrelated (separated by 90 degrees). Thus, the axes miss the clusters, though they lie close to them. Thurstone recognized that clusters might be better defined if axes passed right through them, as in Figure 5. In this so-called "oblique" system of simple structure, the axes themselves are now positively correlated, and this correlation does record a kind of second-order *g,* as Thurstone admitted.

Thus, Jensen argues, *g* must be real. It can't be avoided because it appears both directly in principal components and indirectly as the cause of positive correlation among oblique simple structure axes. Yet Jensen has missed or ignored Thurstone's repeated claim about this indirect form of *g:* it is generally a weak, secondary effect accounting for a small percentage of total variance among all tests. Jensen's argument requires not merely that *g* exist, but that it be, quantitatively, the major source of variance. He writes: "We are forced to infer that *g* is of considerable importance in 'real life' by the fact that *g* constitutes the largest component of total variance in all standard tests of intelligence or IQ."

In sum, there remains a fundamental difference between *g* as the first principal component (Spearman-Burt) and *g* as a second-order correlation of oblique simple structure axes (Thurstone): *g* is usually dominant in the first and very weak in the second—while Jensen's argument requires that it be dominant. Since principal components and oblique simple structure represent two equally valid methods of factor analysis, we are forced to conclude that the dominant *g* required by Jensen is not a fact of nature, but an artifact of choice in methods.

Behind this technicality lies an even deeper error in the identification of *g* with a single quantity defined as "general intelligence." Spearman's *g* is a measure abstracted from correlation coefficients, and the oldest truism in statistics states that correlation does not imply cause (consider the perfect positive correlation between my age and the expansion of the universe during the last five years).[9] Even if dominant *g* were an ineluctable abstraction from the correlation matrix, it still

[9]Times change (though the logic of arguments is eternal). I wrote this in 1980 and originally said "my age and the price of gasoline."

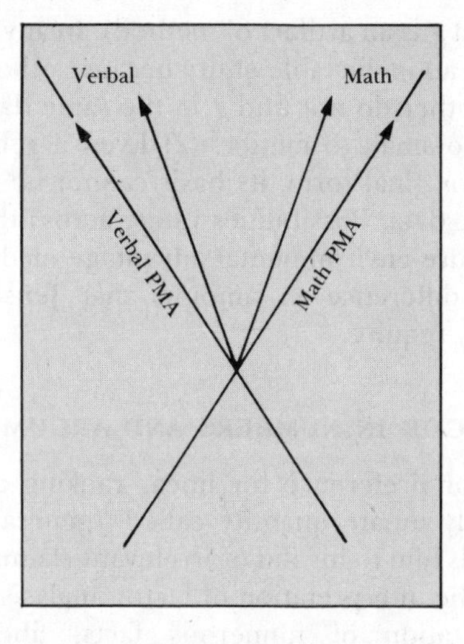

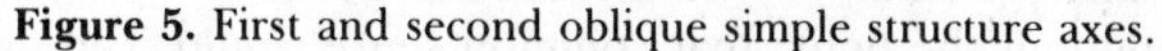
Figure 5. First and second oblique simple structure axes.

wouldn't tell us *why* mental tests tend to be positively correlated. The reason might be largely innate as Jensen assumes and requires—people who do well on one test generally do well on others because they have more inborn intelligence. Or the cause might be largely or totally environmental—people who do well on one test generally do well on others because they had a good education, enough to eat, intellectual stimulation at home, and so forth. The environmental interpretation undermines Jensen's repeated claim (in its vernacular meaning) that whites are more intelligent than blacks as measured by performance on unbiased tests—yet this interpretation is fully consistent with the existence of Spearman's *g*. Jensen's argument suffers a double defeat: First, Spearman's *g* is only one of several valid ways to represent data, not an ineluctable entity. Second, even if *g* were an entity, it could not be identified with any innate property meriting the name "intelligence."

We can conclude: (1) Spearman's *g* is at best one of several ways to summarize sets of data on correlations between mental tests (at worst *g* is an artifact of method). In any case, *g* cannot be viewed as an ineluctable entity because other equally valid techniques either do not find *g* in the same data or find it in quantities too small to matter. (2) Even if *g* be admitted in Spearman's original form, its basis cannot be specified from psychometric data. Possibilities range across the entire spectrum from pure environmental advantage or disadvantage to the inborn difference in amount that Jensen and other hereditarians require.

RIGOR IN NUMBERS AND ARGUMENT

Jensen's prior preference for linear ranking according to a single, largely innate quantity called "general intelligence" not only leads him to invalid or irrelevant claims (the meaning of bias and the interpretation of factor analysis); it also skews his interpretation of numerous facts, liberally studded throughout the book, that common sense would read differently. Jensen, for example, enthusiastically reports a correlation coefficient of brain size and IQ of about 0.3. He doesn't doubt that this correlation records natural selection operating for greater intelligence through larger brains. He regards the strength of this correlation as remarkable because "much of the brain is devoted to noncognitive functions."

Yet at the bottom of the very same page, he records a correlation of equal strength (average of 0.25) between IQ and body stature. This, he doesn't doubt, "almost certainly involves no causal or functional relationship." Jensen is so attached to his preferred scheme of argument that the obvious interpretation of these facts has escaped him—that the weak correlation of IQ and height reflects environmental (largely nutritional) advantages favorable to both and that the correlation of IQ and brain size is a non-causal, indirect consequence of the same relationship since big people have bigger body parts, including brains, arms, and legs (although no one has ever thought of computing a correlation of leg length with brain size. Choice of question is, indeed, a function of expectation).

Other facts, proving (I would think) that environment exerts a powerful influence upon average IQ within groups, are either glossed over or reported in other contexts. I found nothing in Jensen's book more striking than a chart on page 569 showing that children tested in 1972 on the Stanford-Binet and scored according to the 1937 norms have an average IQ of 106.3—a general gain of close to half a standard deviation from the standardized 1937 mean of 100. Within some age classes, the average gain is considerably higher—10.8 points at age 3½, for example (the "Sesame Street" effect, perhaps).

This general gain can hardly be ascribed to genetic causes; it reflects whatever improved literacy, earlier access to information through radio and television, better nutrition, and other cultural changes in a similar vein have wrought in just thirty-five years. When we recognize that the average black-white difference is 15 points, and that gains of up to two thirds this amount have occurred in certain age groups as a result of general changes in environment not specifically directed toward such an end, then why should we be ready to conclude that group differences are ineluctable? I know no fact that so clearly underscores the efficacy of improved standards of living for increasing the so-called "general intelligence" of Americans.

Jensen attempts to cover all these difficulties with the classical ploy of hereditarians: I have the numbers, the rigor, and the objectivity; you have only hopes and emotion. He refers to criticism of testing as "largely emotional, ad hoc, often self-contradictory"; they "convey attitudes and sentiments instead of information." He depicts his own work instead as an "exhaustive review of the empirical research." The ploy often works. Jensen's 800 pages of numbers benumbed many reporters who did not find the basic weaknesses of argument and who therefore assumed, on the more-is-better fallacy, that such length must reflect profundity.

But an argument is only as good as its premises and logic. Jensen may contemptuously dismiss criticisms as "armchair analysis," but the copious citation of numbers cannot salvage an argument grounded on invalid premises. Lord Kelvin proved with numbers more rigorous than any psychometrician has ever derived that the earth could not be more than a few million years old—not enough time for Darwinian evolution. But the numbers rested on a false assumption that heat emanating from the earth's interior reflected the cooling of an initially molten planet, while we know now that this heat arises largely through the decay of radioactive elements.[10]

The computer people have a word for it, one of their most euphonious acronyms—GIGO, or garbage in, garbage out. Jensen's problem is not garbage, but irrelevancy (400 or so pages on bias), and fallacious premises (the equation of Spearman's *g* with intelligence).

Numbers have undoubted powers to beguile and benumb, but critics must probe behind numbers to the character of arguments and the biases that motivate them. Léonce Manouvrier, the leading statistical anthropologist of the late nineteenth century, made this point with feeling when he disproved Paul Broca's claim that the smaller brains of women reflected inferior intelligence:

> Women displayed their talents and diplomas. They also invoked philosophical authorities. But they were opposed by *numbers* unknown to Condorcet or to John Stuart Mill. These numbers fell upon poor women like a sledge hammer, and they were accompanied by commentaries and sarcasms more ferocious than the most misogynist imprecations of certain church fathers. The theologians had asked if women had a soul. Several centuries later, some scientists were ready to refuse them a human intelligence.[11]

[10]See S.J. Gould, "False Premise, Good Science," in *The Flamingo's Smile* (New York: W.W. Norton, 1985), pp. 126–38.

[11]L. Manouvrier, "*Conclusions générales sur l'anthropologie des sexes et application sociales,*" *Revue de l'école d'anthropologie* (1903), 13:405–23.

9

Nurturing Nature

Cat Island in the Bahamas maintains a declining population of about one thousand by slash-and-burn agriculture. Few of the one-room houses have electricity; none has plumbing. The local teacher, a British expatriate, told me that in seven years only one child had managed to win entrance into the two-year program at the College of the Bahamas in Nassau—and that she had flunked out. When I asked why, he gave a perfectly obvious and reasonable answer: how can Cat Island children maintain any interest or time for studies? They come home late in the afternoon; they have to haul water, care for the goats, help to prepare food. After dinner, they have no place (or light) for doing homework. I nodded in evident agreement, but his next statement startled me (this, I should add, was a casual barroom conversation; he knew me only as a peculiar snail collector, not as author of *The Mismeasure of Man*). We now *know,* he said, that only 20 percent of mental ability is environmental; 80 percent is inherited, so these immediate factors can explain, at most, one-fifth of the underachievement. The rest must be genetic, probably caused (he opined) by inbreeding among the few families that inhabit Cat Island.

Lee Kuan Yew, the prime minister of Singapore, recently raised a furor in that distant land by suggesting that the ge-

A review of *Not in Our Genes: Biology, Ideology and Human Nature* by R.C. Lewontin, Steven Rose, and Leon J. Kamin.

CHARLES DARWIN

netic stock of his nation is about to plummet. He studied his census figures and noted a trend common to all developed nations: highly educated women are having fewer children than women with little schooling. Although this fact usually (and correctly) inspires no action beyond a call for more education (both for its intrinsic merits and for its salutary impact upon population size), Lee gave the argument a discredited eugenic twist that has not been heard for the past half century or so: uneducated women are genetically inferior in intelligence and their likeminded offspring will swamp the smaller pool of intrinsically bright children born to educated parents. Lee acknowledged that environment and upbringing can influence both access and success in education, but we now *know*, he continued, that 80 percent of intelligence is fixed by inheritance, and only 20 percent malleable by circumstance. "A person's performance," Lee stated, "depends on nature and nurture. There is increasing evidence that nature, or what is inherited, is the greater determinant of a person's performance than nurture (or education and environment). . . . The conclusion the researchers draw is that 80 percent is nature, or inherited, and 20 percent the differences from different environment and upbringing."

The fallacies of this and other hereditarian arguments about complex human social behaviors have been so thoroughly rehearsed that scholars might be tempted to treat any new discussion with undisguised boredom. In the case of IQ, estimates of heritability are a confusing mess, ranging from the notorious 80 percent, long cited by Jensen and based originally upon the faked research data of Sir Cyril Burt, to Leon Kamin's argument that existing evidence does not preclude an actual value of zero. In any case, and much more importantly, heritability, as a technical term, simply doesn't bear its vernacular meaning of "inevitability"—the essential component of the argument's public use, as my two initial examples indicate. Heritability is not a measure of flexibility, but a statement about how much variation for traits within populations can be attributed to genetic differences among individuals. Some visual impairments are nearly 100 percent heritable, but easily corrected with a pair of eyeglasses.

Whatever its status on our campuses (where confusion and obfuscation are, as usual, by no means absent), the crudest, discredited hereditarian argument about IQ still influences and restricts the lives of millions. So long as teachers on tiny islands and prime ministers of major nations act upon their belief that 80 percent of intelligence is fixed in the genes, human potential will be sacrificed on an altar of misunderstanding. Biological determinism is, fundamentally, a theory of limits.

For these reasons, *Not in Our Genes* is an important and timely book, for it not only exposes the fallacies of biological determinism (a field perhaps well enough plowed)[1] but also presents a positive view of human behavior that could propel us past the stupefying sterility of nature-nurture arguments. A proper understanding of biology and culture both affirms the great importance of biology in human behavior and also explains why biology makes us free. The old equation of biology with restriction, with the inherent (as opposed to the malleable) side of the false dichotomy between nature and nurture, rests upon errors of thinking as old as Western culture itself. The critics of biological determinism do not uphold the equally fallacious (and equally cruel and restrictive) view that human culture cancels biology. Biological determinism has limited the lives of millions by misidentifying their socioeconomic disadvantages as inborn deficiencies, but cultural determinism can be just as cruel in attributing severe congenital diseases, autism for example, to psychobabble about too much parental love, or too little.

As a contribution to the ever troubling and important issue of biological determinism, *Not in Our Genes* possesses two special strengths. We must first praise the authors' unusually honest self-analysis of the reasons for their concern. This frankness can lead us beyond the conventional set of self-serving myths to a better understanding of how good scientists work. Richard Lewontin, Steven Rose, and Leon Kamin bring a comprehensive range of expertise to their subject. Lewontin is a population geneticist and author of a recent book on the causes and meaning of human diversity; Rose works in the neurosciences and has written many fine analyses of the relationship between brain structure and human behavior; Kamin,

[1]For example, A. Chase, *The Legacy of Malthus* (New York: Knopf, 1977); S. Chorover, *From Genesis to Genocide* (Cambridge: MIT Press, 1979); Stephen Jay Gould, *The Mismeasure of Man* (New York: W.W. Norton, 1981); Leon J. Kamin, *The Science and Politics of IQ* (Lawrence Erlbaum Associates, 1974); K. Ludmerer, *Genetics and American Society* (Baltimore: Johns Hopkins University Press, 1972).

a psychologist, first exposed the fakery of Sir Cyril Burt and wrote an important account of the history and meaning of IQ tests.[2]

Amid this diversity, the authors share (with this reviewer, I must add in good conscience) a definite and frankly stated perspective on biological determinism in particular and on the social function of science in general. They write in their preface:

> Each of us has been engaged for much of this time in research, writing, speaking, teaching, and public political activity in opposition to the oppressive forms in which determinist ideology manifests itself. We share a commitment to the prospect of the creation of a more socially just—a socialist—society. And we recognize that a critical science is an integral part of the struggle to create that society, just as we also believe that the social function of much of today's science is to hinder the creation of that society by acting to preserve the interests of the dominant class, gender, and race.

The traditional and unthinking response to such frankness by scientists is outright dismissal of any subsequent statement on grounds of *prima facie* bias. After all, isn't science supposed to be a cool, passionless, absolutely objective exploration of an external reality? As T.H. Huxley said in his famous letter to Charles Kingsley,[3] so often taken out of context and misinterpreted in just this naive light,

> Sit down before fact as a little child, be prepared to give up every preconceived notion, follow humbly wherever and to whatever abyss nature leads, or you shall learn nothing.

[2]R.C. Lewontin, *Human Diversity* (W.H. Freeman, 1984); Steven Rose, *The Conscious Brain* (New York: Knopf, 1973); Leon J. Kamin, *The Science and Politics of IQ* (Lawrence Erlbaum Associates, 1974).

[3]See my essay, "Moon, Mann and Otto," reprinted in *Hen's Teeth and Horse's Toes* (New York: W.W. Norton, 1983).

But we scientists are no different from anyone else. We are passionate human beings, enmeshed in a web of personal and social circumstances. Our field does recognize canons of procedure designed to give nature the long shot of asserting herself in the face of such biases, but unless scientists understand their hopes and engage in vigorous self-scrutiny, they will not be able to sort unacknowledged preference from nature's weak and imperfect message. As Herbert Butterfield wrote in his great essay, *The Whig Interpretation of History*:

> The historian may be cynical with Gibbon or sentimental with Carlyle; he may have religious ardor or he may be a humanist. . . . It is not a sin in a historian to introduce a personal bias that can be recognized and discounted. The sin in historical composition is the organization of the story in such a way that bias cannot be recognized.

An overtly expressed political commitment does not debar a scientist from viewing nature accurately—if only because no honest scientist or effective political activist would be foolish enough to advance a program in evident discord with the world as we find it. Many facts of nature are decidedly unpleasant—the certainty of our bodily death prominently among them—but no social system fails to incorporate these data (despite a plethora of palliations, from reincarnation to resurrection, advocated by many cultures).

The proper relationship between nature and the personal and social lives of scientists lies in an admittedly simplified and venerable distinction long advocated by logical positivism—the difference between "context of discovery" and "context of justification." If you wish to know why Lewontin and not geneticist X reached a certain conclusion or why he did so in 1984 and not in 1944—all questions about context of discovery—then examine psychohistory and socioeconomic circumstances. But "truth value"—or context of justification—is a different matter. People reach conclusions for the damnedest of peculiar reasons: pure guesses inspired by poetry dimly remembered during a dream have sometimes turned out to be true, while conclusions meticulously reached by conscious and repeated experiment may be wrong.

Leftist scientists are more likely to combat biological determinism just as rightists tend to favor this quintessential justification of the status quo as intractable biology; the correlations are not accidental. But let us not be so disrespectful of thought that we dismiss the logic of arguments as nothing but an inevitable reflection of biases—a confusion of context of discovery with context of justification. If we thought that biological determinism was pernicious but correct, we would live with it as we cope with the fact of our impending death. We have campaigned vigorously against this doctrine because we regard determinist arguments primarily as bad biology—and only then as devices used to support dubious politics.

Not in Our Genes is an analysis of determinist argument from a definite point of view; it is not a political diatribe. It begins with several chapters on the historical origin and social utility of claims that inequalities among races, classes, and sexes reflect the differential genetic worthiness of individuals so sorted. Subsequent chapters analyze the details of major contemporary arguments in the determinist mode: IQ, patriarchy, the attribution of social pathology among the poor and dispossessed to diseased brains, schizophrenia (where Kamin tries to apply the same kind of detailed reanalysis of case studies that he used in his successful debunking of IQ, and finds much superficial shoddiness and inconsistency, but not, I think, fatal and debilitating flaws), and sociobiology. The last chapter, "New Biology versus Old Ideology," presents a positive view of a proper and inextricable relationship between biology and culture.

The second major strength of *Not in Our Genes* lies in its attempt to progress beyond debunking by providing a useful model of how biology creates and interacts with culture (which then creates and interacts with biology). Lewontin, who ought to know since he serves as a volunteer fireman in southern Vermont, laments that fighting biological determinism is like putting out fires. Every time you extinguish one, another starts somewhere else. No sooner have you discredited Carleton

Coon's theory of the parallel and separate origins of human races from different stocks of *Homo erectus* (with blacks making the last transition and therefore still lagging behind) than Robert Ardrey writes a colorful book about the origin of human violence in the territoriality of carnivorous *Australopithecus,* the killer ape. So you show that *Australopithecus* was predominantly a herbivore. Then William Shockley argues that IQ declines among American blacks in direct proportion to the percent of their African heritage (and also proposes that we pay a voluntary sterilization bonus scaled to the extent of this measured deficit). By now you're exhausted; you never want to slide down that damned fire pole again. But the authors of *Not in Our Genes* breathe deeply, and attempt a positive formulation.

The straw man set up to caricature biological determinism is cultural determinism or the tabula rasa in its pure form. Although biological determinists often like to intimate, for rhetorical effect, that their opponents hold such a view, no serious student of human behavior denies the potent influence of evolved biology upon our cultural lives. Our struggle is to figure out how biology affects us, not whether it does.

The first level of more sophisticated argument that goes beyond crude nature-nurture dichotomies is "interactionism"—the idea that everything we do is influenced by both biology and culture, and that our task is to divide the totality into a measured percentage due to each. In fact, this kind of interactionism is the position of most biological determinists, who love to argue that they are not crude 100 percenters of pure naturism (of course they are not: just as no one is quite so stupid as to nullify biology completely, so too does no one deny some flexibility in the translation of genes into complex behaviors). Biological determinists hide behind the screen of interactionism, complain bitterly that they have been maligned, and that they do, after all, acknowledge the importance and independence of culture. They then allot the percentages so that genes control what really matters—80 percent determinism, after all, is usually good enough for the cause. On this model, antideterminists are the folks who do the parceling out differently and grant only a few percent to the genes.

But, as Lewontin, Rose, and Kamin emphasize in the main

theme of their book, interactionism is also based on deep fallacies and cultural biases that play into the hands of biological determinism. This mechanical brand of interactionism still separates biology and culture; it still views genes as primary, deep, and real, and culture as superficial and superimposed. Genes are our inherited essence, culture the epiphenomenal tinkering.

The chief fallacy, they argue (I think correctly), is reductionism—the style of thinking associated with Descartes and the bourgeois revolution, with its emphasis on individuality and the analysis of wholes in terms of the underlying properties of their parts.

We must, they argue, go beyond reductionism to a holistic recognition that biology and culture interpenetrate in an inextricable manner. One is not given, and the other built upon it. Although stomping dinosaurs cannot make continents drift, organisms do create and shape their environment; they are not billiard balls passively buffeted about by the pool cues of natural selection. Individuals are not real and primary, with collectivities (including societies and cultures) merely constructed from their accumulated properties. Cultures make individuals too; neither comes first, neither is more basic. You can't add up the attributes of individuals and derive a culture from them.

Thus, we cannot factor a complex social situation into so much biology on one side, and so much culture on the other. We must seek to understand the emergent and irreducible properties arising from an inextricable interpenetration of genes and environments. In short, we must use what so many great thinkers call, but American fashion dismisses as political rhetoric from the other side, a dialectical approach.

Dialectical thinking should be taken more seriously by Western scholars, not discarded because some nations of the second world have constructed a cardboard version as an official political doctrine. The issues that it raises are, in another form, the crucial questions of reductionism versus holism, now so much under discussion throughout biology (where reductionist accounts have reached their limits and further progress demands new approaches to process existing data, not only an accumulation of more information).

When presented as guidelines for a philosophy of change, not as dogmatic precepts true by fiat, the three classical laws of dialectics embody a holistic vision that views change as interaction among components of complete systems, and sees the components themselves not as a priori entities, but as both products of and inputs to the system. Thus the law of "interpenetrating opposites" records the inextricable interdependence of components: the "transformation of quantity to quality" defends a systems-based view of change that translates incremental inputs into alterations of state; and the "negation of negation" describes the direction given to history because complex systems cannot revert exactly to previous states.

Groucho Marx caught the spirit of academic pettiness well when he delivered his inaugural address in song as president of Darwin (or was it Huxley) College in *Horsefeathers:* "Whatever it is, I'm against it." By contrast, Lewontin, Rose, and Kamin have entered a prime area of academic debunking and emerged with a positive program. Indeed, they are calling for no less than a revolution in philosophy. They are also not unmindful of that oldest chestnut in the Marxist pantheon (Karl this time), the last thesis on Feuerbach: philosophers thus far have only interpreted the world in various ways; the point, however, is to change it.

IV

Four Biologists

10

Triumph of a Naturalist

It would be churlish to argue that Watson and Crick's elucidation of the double helical structure of DNA in 1953 was anything less than one of the great scientific achievements of modern history. Yet, in a curious way, this discovery differed from other revolutionary events in science by its doubly conservative nature. First, it seemed to confirm the canonical view of genetical systems as rows of beads (genes) on strings (chromosomes), with its implication that evolution proceeds slowly and gradually, based ultimately on the generation of new genetic variation (mutation) by pinpoint changes within the beads (substitution of one nucleic acid base for another yielding a different amino acid in the translated protein).

Second, Watson and Crick's discovery represented the greatest modern triumph of the standard methodology that has ruled orthodox science ever since Descartes: reductionism, with its particular claim (in this case) that the complex forms of completed organisms could be explained, ultimately, as the translated products of an inherited program coded by a simple mechanism of four bases variously arranged in groups of three. Had not Watson built his double helix as a tinker-toy machine based on the sizes and fits of constituent parts? And had not Crick, soon afterward, proclaimed the "central dogma" (his words) of the new biology—that DNA makes RNA

A review of *A Feeling for the Organism: The Life and Work of Barbara McClintock* by Evelyn Fox Keller.

BARBARA McCLINTOCK

and RNA makes protein, in a one-way flow of information, a unidirectional process of mechanical construction?

It is a credit to the power of Watson and Crick's model and to the fruitfulness of good science in general that, thirty years later, this Cartesian view of molecular genetics has been superseded, as a second revolution transmutes our view of inheritance and development. The genome, a cell's compendium of genetic information, is not a stationary set of beads on strings, subject to change only by substituting one bead for another. The genome is fluid and mobile, changing constantly in quality and quantity, and replete with hierarchical systems of regulation and control.

Genes come in pieces, and the shuffling of their segments can produce new combinations. Some genes can excise themselves from a chromosome and move to other locations in the genome; if these "transposable elements" operate as regulators to turn adjacent genes on and off, their movement to new places (near different genes) can have major effects on the timing and control of development. Other genes make copies of themselves, and these identical daughters can then either reside next to their parent or move to other chromosomes. In this way, hundreds or thousands of copies of the same gene may be repeated within an organism's genetic program. The multiple copies of this "gene family" may then diverge in function, thus providing a solution to the old conundrum of how anything novel can evolve if all genes make products necessary for an organism's construction and well-being. (Original copies may continue to make the required product, while new copies are "free" to alter and experiment.)

If these processes replace a static with a more mobile genome subject to rapid and profound rearrangement, the central dogma with its one-way flow of information from code to product has also been breached. A substance called "reverse transcriptase" can read RNA into DNA and insert new material into genetic programs by running backward along the supposedly one-way street of the central dogma. A class of objects, called retroviruses (and including the cause of AIDS), uses this backward path, placing new material into chromosomal DNA from the outside. In short, a set of new themes—mobility, rearrangement, regulation, and interaction—has transformed our view of genomes from stable and linear arrays, altered piece by piece and shielded from any interaction with their products, to fluid systems with potential for rapid reorganization and extensive feedback from their own products and other sources of RNA. The implications for embryology and evolution are profound, and largely unexplored.

Barbara McClintock is the godparent and instigator of this second revolution. Her discovery of transposable elements in maize—first presented in the early 1950s before her field had any language to express such a heterodox idea—was, in retrospect, the beginning of modern molecular genetics. She suffered the fate of many pioneers—incomprehension and bewilderment from most colleagues who could not read her maps of terra incognita. But by tenacity, the blessings of long life, and continuous fruitful activity, she has avoided the maudlin ending of most tales in the annals of exploration, and has lived to savor her triumph in the midst of an active career. Now in her eighties, and committed as ever to research on maize in her laboratories at Cold Spring Harbor, she has won every major award that science and an adoring public can bestow, from MacArthur laureate to Nobel prize. And, as a supreme irony, this intensely private woman, who has worked all her life for personal and intellectual reward, can only view such recognition as a bother and impediment. (Most of us, so honored, feel compelled to make some public comment to the same, selfless effect, but we love the accolade and bask in the notoriety; I believe that Barbara McClintock may be unique in truly feeling more discomfort than vindication.)

Such heroic tales are the stuff of simplistic mythology, and McClintock's catapult into public recognition has fostered vulgar versions of what she did, thereby obscuring a more subtle story and, in a perverse if unintended way, degrading McClintock's formidable achievements. The vulgarized accounts try to use her as an exemplar for one of two archetypical stories in the sociology of science, either (1) the woman in science, a brilliant mind rejected by prejudice against the color or sex of the body housing it, or (2) the maverick genius who, despite heroic efforts, obtains no recognition because colleagues simply cannot hear a different drummer. The story is never so simple, never a clear-cut contrast of unblemished individual genius and benighted establishment. Just as McClintock's work helped to break the central dogma and establish interaction between code and product, so must the complex tale of her long rejection be cast as an interplay between her own idiosyncracies and the reactions of her colleagues.

The strength of Keller's fine book lies in her successful attempt to avoid the myths and capture the subtleties, thereby providing a rare and deep understanding of a troubling, fascinating, and general tale in the history of science—initial rejection (or, more frustratingly, simple incomprehension) of great insights.

Yet, in a kind of Catch-22, the uniqueness that makes McClintock such a fascinating subject, and that prevents her from serving as a prototype for any standard tale, has also limited Keller's achievement for reasons utterly beyond her control. If McClintock has prevailed because a remarkable ability to work for her own satisfactions rather than for approbation of peers kept her going with equanimity and fortitude through all the years of rejection, then the same love of privacy makes her a most unforthcoming subject for a biography. She provided Keller with precious little material of the sort that makes a good story—a few factual tidbits here and there (but little beyond the public record) and almost no account of her feelings and motives.

The first theme, the woman in science, fails to explain her long years of intellectual loneliness after the discovery of "jumping genes" in maize. I do not say, of course, that she suffered no prejudice on the basis of sex; such discrimination was pervasive and affected every stage of her developing career to her practical disadvantage. Yet it would denigrate McClintock's remarkable achievements to attribute her experience of isolation, a key event in her intellectual history, to blind prejudice that she could not control. For Barbara McClintock, in her distinctive and personal way, had overcome these pervasive prejudices by the acknowledged brilliance of her work long before she discovered transposable elements in maize.

Great achievements often wipe out public memory and restart a scientist's professional clock. Darwin, for example, pursued a successful career as a geologist before he published a word on the origin of species, but how many among his admirers and denigrators know that he solved the problem of the origin of coral atolls? Since McClintock has been so honored for the jumping genes that ushered in the second revolution in molecular genetics, many commentators do not realize that transposable elements were not the discovery of a brash young investigator, but a further step built upon a recognized and distinguished career.

Since no one can look consciously for the unexpected, McClintock found transposable elements while probing another question then largely ignored by geneticists, but of central importance to the study of embryology, and particularly suited to her talents. She wanted to learn how some genes affect the timing of development by regulating activity of other genes that build parts of the body. Maize is particularly well suited for these studies because differences in timing often translate into clear effects in the color of husks and kernels. Her reductionist colleagues, who avoided such complex creatures and tried to get closer to molecules by manipulating the simplest of unicellular organisms, were not then studying problems of regulation—though many molecular geneticists do today because the issue is so fundamental to what Aristotle recognized as the central problem of biology: the development of organic form. McClintock found regulatory elements in maize and then, incidentally while probing their inheritance and expression, discovered that they move.

McClintock could study the difficult problem of regulation successfully because she had already spent a career as one of America's most distinguished students of cytogenetics—the branch of biology that investigates the physical basis of inheritance by linking observed genetic patterns with the structure of chromosomes and other components of the cell. McClintock had ushered in the modern era of maize genetics in a series of careful studies that developed techniques for naming, visualizing, and characterizing the ten pairs of chromosomes that carry the DNA of corn. She also performed a series of classic experiments in cytogenetics that established the physical ground for basic principles of heredity. Most famous was her proof, published with Harriet Creighton in 1931, that chromosomes really do carry genetic programs in the linear order that traditional breeding experiments with Mendelian pedigrees had indicated. (She did this by studying the process of "crossing over," when chromosomes paired during meiosis—the cellular divisions that form eggs and sperm—exchange genes. She proved that this genetic crisscross corresponds exactly with the breaking off and mutual exchange of chromosomal segments.)

For this elegant and pathbreaking work, and in the face of continual impediments raised by prejudices against women in science, McClintock was abundantly recognized and honored by her colleagues. She served as vice-president of the Genetics Society of America in 1939, and as president in 1945. In 1944 she reached the pinnacle of peer recognition in American science and became the third woman ever elected to the National Academy of Sciences. At that time, she wrote to Tracy Sonneborn in the only contemporary, explicit reference she ever made (in Keller's citation) to the problems of women scientists:

> It was both thoughtful and generous of you to write me as you did concerning the National Academy. I must admit I was stunned. Jews, women and Negroes are accustomed to discrimination and don't expect much. I am not a feminist, but I am always gratified when illogical barriers are broken. . . . It helps all of us.

McClintock surely suffered from all the prejudices, subtle and overt, directed against women in science, but she overcame them by dint of personal genius and an awesome inner strength that few of us can hope to possess. She will not serve as an exemplar of this troubling theme, and it will not explain the primary event of her public life—the stony silence that accompanied her most important discovery of transposable elements.

The second theme—genius so far ahead of her time that no one can understand—also contains a partial truth but will not suffice because it foists all explanation upon external reception beyond McClintock's control. Her transition from peerless scientist to pariah owed as much to her personal style as to any inability of colleagues to grasp a radical new idea.

McClintock has always worked for herself and in her own way, never tailoring her efforts to win acceptance, or even to promote understanding among those who might need a little extra prodding or clarity to appreciate an unconventional notion. She lost the only regular academic job she ever held (at the University of Missouri, where women could, and occasionally did, win promotion) largely as a consequence of her personal preferences (her disinclination to teach formal classes, and her contempt for what academic departments call "good citizenship"—primarily a euphemism for submission to myriad, meaningless hours of soul-sapping committee work). Some of the idiosyncracies that Keller describes are certainly mild, and we must therefore accept her argument that the combination of woman and maverick, in synergism, ultimately led to McClintock's dismissal. (One day, for example, arriving at her lab without keys, she climbed up the side of the building and crawled in through a window, all in the unsuspected presence of a hidden photographer. But what else could she do? Who wouldn't try acrobatics in such a frustrating situation? I certainly would, and have.)

More importantly, she never did all she could to promote her chances in the admittedly tough battle for acceptance and understanding of transposable elements. She planned a minor campaign, writing an introductory paper and presenting a seminar to key people at Cold Spring Harbor. But when these initial forays met with little response and general incomprehension, she folded her tents. With her usual fortitude and self-reliance (though not, of course, without bitter disappointment), and invoking her own version of the immemorial response "bugger them," she pressed on in her own way, knowing that she was right and that the rest of the world would eventually catch up. She published most of her subsequent work in the annual reports of her laboratory, surely an inauspicious place to launch a scientific revolution. I have read her main papers and they are, to put it mildly, tough slogging. With their unrelenting passive voice, and their compression of complex reasoning and experiment into single paragraphs, they are marvels of their genre but not models of optimal communication.

Keller argues, I think correctly, that a third reason eclipses gender and collegial obtuseness as a primary explanation for McClintock's years as *vox clamantis in deserto*—her unconventional style of scientific thinking. Her different inspiration transcends mere empirical contention and embraces a mode of reasoning quite foreign to the procedures of most experimental science.

First, and more general, McClintock does not follow the style of logical and sequential thinking often taken as a canonical mode of reasoning in science. She works by a kind of global, intuitive insight. If she is stuck on a problem, she will not set it out in rigorous order, write down the deduced consequences and work her way through step by step, but will take a long walk or sit down in the woods and try to think of something else, utterly confident that a solution will eventually come to her *in extenso.* This procedure makes scientists suspicious and has often led colleagues to label her as a "mystic" in the pejorative, not appreciative, sense.

Nothing could be more inappropriate than the word "mystic" applied to this style of reasoning. It is a common procedure for some people, though perhaps rare (and certainly not generally appreciated) in science. It is neither mystical nor, in another vulgar misrepresentation, feminine as opposed to masculine in character. We dub this style mysterious because we have neither good words nor concepts in our largely linear language to express such a modality. I am particularly sensitive to its denigration because I happen to work in the same way. (I am hopeless at deductive sequencing and can never work out the simplest Agatha Christie or Sherlock Holmes plot—the best stereotypical representations of this conventional mode. I never scored particularly well on so-called objective tests of intelligence because they stress logical reasoning and do not capture this style of simultaneous integration of many pieces into single structures. The difference, of course, between McClintock and me is that she is a genius who can depend upon integrative insight for the solution to major scientific problems; I can only be sure that the "correct" outline for an article will eventually come to me *in toto.*)

I think that the best description of this style was presented not by a psychologist or neurologist, but by another mystery writer, Dorothy Sayers, who, I am convinced, worked this way herself and established Lord Peter Wimsey as a conscious antidote to the Sherlock Holmes tradition of logical deduction. Now Wimsey was no intellectual slouch, and (in his impeccable upper-class style) was certainly not a fuzzy romantic mystic either. But he solved his cases by integrative insight (and these I usually do figure out). In the first Wimsey novel, *Whose Body,* Sayers describes the process explicitly:

> And then it happened—the thing he had been half-consciously expecting. It happened suddenly, surely, as unmistakably, as sunrise. He remembered—not one thing, nor another thing, nor a logical succession of things, but everything—the whole thing, perfect, complete, in all its dimensions as it were and instantaneously; as if he stood outside the world and saw it suspended in an infinitely dimensional space. He no longer needed to reason about it, or even to think about it. He knew it.

Second, and more specifically, McClintock practices a style of biology quite foreign to the norms of molecular biology and genetics (and often denigrated, sometimes explicitly, by leaders of these professions—though less so now than before, as appreciation and recognition mount). Keller captures this style in her well-chosen title—"A Feeling for the Organism."

The experimental sciences (like molecular biology) generally work in the reductionistic mode, trying to establish simple and linear chains of cause and effect. They prefer explanations rising from the lowest level of molecules and their physicochemical properties. To reach this "basic" level, they work with the simplest organisms and try consciously to avoid the individuality of any particular creature. They concentrate instead on the repeatable properties of large groups (so that a clone of bacteria becomes the analogue of a population of atoms with no individuality by definition).

This style has enjoyed remarkable success in the history of biology, though I believe that it has now reached definite limits in attempting to understand genetic systems and their complex interactions with the developing form of organisms. It must also be admitted that, although McClintock first found transposable elements with her different manner of working, the discovery that proved her right and elevated her to heights of peer and public recognition came from molecular biologists working with simple unicellular organisms as physical objects.

McClintock's style is not uncommon; it just isn't widely used in her own subfield of biology. In fact, McClintock's mode is the procedure of my own discipline—evolutionary and taxonomic biology. (As in most scientific disciplines, and for the same bad reasons, we are—and have been ever since Linnaeus—predominantly male. So taxonomic thinking cannot be interpreted as a feminine style either.) We work directly with complex organisms and their interaction with each other and their physical environment in growth and adult life. We accept the individuality of each organism as fundamentally irreducible, as the definition of biology's uniqueness and complexity. (This individuality is, for example, the source of Darwinian evolution since natural selection cannot operate unless populations present a wide spectrum of variation among their constituent members.)

McClintock chose an organism that most molecular geneticists would shun as recalcitrant and hopelessly complex for discovering anything fundamental about genetic systems. Maize grows only one generation per year, and its many parts are subject to a baffling array of noncoded modifications that respond to local environments of growth. (Bacteria may divide in twenty minutes. Populations of billions, with no discernible individuality—save for valued new mutuants that can then be cloned by billions themselves—are readily generated.) But McClintock has always believed that scientists must follow the peculiarities of individuals, not the mass properties of millions. She is a true taxonomic biologist, a naturalist not a mystic, working in a field unfamiliar with (and often alienated from) this approach. She said to Keller that one must understand

> how it grows, understand its parts, understand when something is going wrong with it. [An organism] isn't

just a piece of plastic, it's something that is constantly being affected by the environment, constantly showing attributes or disabilities in its growth. . . . No two plants are exactly alike. . . . I start with the seedling, and I don't want to leave it. I don't feel I really know the story if I don't watch the plant all the way along. So I know every plant in the field. I know them intimately, and I find it a great pleasure to know them.

I see a happy lesson in McClintock's story, and in the triumph of her unconventional style. She chose to work as a naturalist with a complex organism that most colleagues wouldn't touch. With maize, she could study basic problems that bacteria do not well exemplify—genetic regulation of timing in growth and morphogenesis, for example. But when she made her unanticipated and greatest discovery of transposable elements, confirmation and generalization required the different procedures of reductionistic molecular genetics. Biology is a unity, and we will not solve Aristotle's dilemma of morphogenesis, the origin and development of organic form, until we marry the distinctive styles of natural history and reductionistic experiment. Barbara McClintock, with her "feeling for the organism" and her uncanny ability to perform the most rigorous and elegant experiments, points the way better than any other scientist I know.

11

Thwarted Genius

Like so many other American biologists, I have spent several summers working at the Marine Biological Laboratory in Woods Hole, and my affection for the place runs deep. During my first summer there, I would often wander about very late at night (or early in the morning, since the library never closes and the joy of research easily fuses night with coming day). I would study the numerous plaques, medals, and photographs hanging here and there, dedicated to the great scientists who had worked at Woods Hole. One particularly caught my attention, a photo hanging in the main reading room, labeled Ernest Everett Just. The man depicted was singularly handsome, with a pervasive look of sadness that touched me across half a century. It differed from all others in a way that shouldn't matter but always has in America—E.E. Just was black. I became fascinated with the man, read all I could by and about him, realized how interesting and ambiguous a story his life and work had been, and have wondered ever since why nobody had written his biography.

I am now delighted to report that MIT historian of science Kenneth R. Manning has published a detailed biography of Just, and that it is among the finest accounts of a scientist's life that I have ever read. Manning has chosen to write an institutional history of science. He says relatively little about Just's

A review of *Black Apollo of Science: The Life of Ernest Everett Just* by Kenneth R. Manning.

E. E. JUST

biological research, and less about the biological theories of his time.[1] He concentrates instead on Just's relationship to institutions: Howard University where he taught, the laboratories in Woods Hole and Europe where he worked, and especially the foundations where he relentlessly sought funding for research. Since mavericks and unusual situations best reveal the normal character of institutions, Just's story, fascinating in

[1]See my essay on Just's biological theories, "Just in the Middle" reprinted in *The Flamingo's Smile* (New York: W.W. Norton, 1985).

itself as human drama, incisively reveals the practice of science, and the structure of its financing, in early twentieth-century America. It also serves as a guide to the racial attitudes of American scientists, a group who considered themselves (and may even have been) among the most enlightened in prejudiced America.

E.E. Just was born in Charleston, South Carolina, in 1883. His grandfather, a freed slave probably fathered by his former owner, was a skilled wharf builder and a leader of Charleston's black community. His father was an alcoholic who died young, and Just was raised by his mother, a strong and determined woman with unswerving religious and educational commitments. In an unusual move, Just was sent north for schooling, first to Kimball Union Academy in Vermont, thence to Dartmouth, where he began as a student of English and classics, switched to biology, and graduated as the only *magna cum laude* of 1907 (there were no *summas* that year).

Today, such a promising black scholar would be wooed assiduously by every major university in America. For Just, as Manning notes, possibilities were much more restricted: "An educated black had two options, both limited: he could either teach or preach—and only among blacks." So Just went to Howard and stayed there all his life, with zero institutional mobility, despite superb research and copious publication. Howard, with no graduate program and crushing responsibilities in teaching and administration, was singularly unsuited for the pure research in experimental embryology that Just longed to pursue. Just therefore contacted Frank R. Lillie, head of the Marine Biological Laboratory at Woods Hole, and he obtained a summer's assistantship there in 1909. The two men immediately sensed their mutual affinity. Both were quiet and conservative in temperament, sticklers for absolute cleanliness and careful procedure in the laboratory, and committed to a nonreductionist approach to embryological problems. Just spent two decades of summers working for Lillie at Woods Hole, and wrote most of his eighty scientific papers and two books there.

But the racism that usually lay under the surface (but flared up explicitly the one summer that Just tried to bring his wife and children along, only to have them ignored by everyone) and, above all, the crushing paternalism that emerged from even the friendliest quarters, even (if not especially) from Lillie, eventually drove this proud man to anger and distraction. He had beat the bushes of funding for years, with a good deal of success, considering the obvious obstacles (he had even managed to find some support during the Depression when many white Ph.D.'s were on the bread lines). Consequently, he was able to visit Europe and pursue research in German laboratories and French and Italian marine stations. Here a new world opened. Finally, in mid-life, he was accepted simply for the quality of his work and encouraged without impediment. Just angrily rejected America and embraced a new life in Europe.

He became so attached to his romanticized view of European culture that he either misunderstood or chose to ignore the encroaching reality of fascism. He was working at a marine station when the Nazis took France in 1940. Bitter and confused, Just came back—where else?—to Howard, where he died of pancreatic cancer only a year later, in October 1941.

Just's biological work lay firmly in the great classical tradition of experimental embryology that made Woods Hole a center of American biology early in the century. His experiments were noted for their care and meticulousness of design. His empirical work focused on problems of fertilization, where he made several notable contributions. (In his very first paper, for example, he showed that, at least in some important marine invertebrates, the plane of first cleavage corresponds to the point of entry of the sperm. Since he could also demonstrate that the egg's surface was "equipotential"—for the sperm had no favored spot of entry—the plane of cleavage must be determined exogenously by the point of fertilization, not by any prearrangement of material within the egg. This may not strike most readers, even the biologically sophisticated, as a hot issue today, but Just's experiment spoke to the major debate within embryological circles at the time—a reincarnation of the oldest issue in biology, epigenesis versus preformation.)

As his work on fertilization progressed, Just became convinced that the cell surface cannot be viewed as a passive boundary. The surface, he argued, is an active determinant of essential cellular processes, endowed with far greater complexity of design and chemistry than usually attributed to simple boundaries. This conclusion led Just to a holistic view of biological organization, a powerful and sensible midway position between the outright vitalism of some traditional European biologists (who believed in a "special spark" inherent in life alone and were more than just vaguely mystical) and the mechanistic reductionism of some zealous "modernists" (who believed that biology was nothing but physics and chemistry, explainable in terms of basic constituents, and requiring no concept of higher levels of organization). Just's work, particularly after his European exile, thus took on a more philosophical tone, culminating in his general book of 1939 with its deceptively pedestrian title, *The Biology of the Cell Surface.*

This short account of Just's research should suggest the special challenge that he posed even (or especially) to whites who considered themselves enlightened in their racial views. As Manning notes so well, biological liberals of the time were quite prepared to accept black scientists who, like George Washington Carver, allied themselves with Booker T. Washington's doctrine of slow and humble self-help and devoted their lives to practical efforts in aiding black farmers to find more uses for peanuts. From this perspective, Just was a threatening anomaly. He wanted to do theoretical research at the highest levels of abstraction and detail—and he performed well. He was a proud and cultured man who loved good wine and good clothes, white women, opera, and the classics—though, as a practical man as well, he also cultivated "humility," even shuffling, when the situation demanded compromise, at least until he could stand the charade no longer. "The white public," Manning writes, "felt more comfortable reading about a black scientist wearing dirty aprons and doing manual labor than about one dressed in elegant suits and preparing high-powered articles for serious scientific journals."

Just's career therefore posed a special challenge to the institutions of science in general and to the institutions of funding in particular. Just was a zealous self-promoter (how else could he have survived at all?), and his assiduous efforts brought him into contact with nearly everyone who mattered in American biology and philanthropy.

Reactions to Just were as varied as the men and institutions he approached, but they were dominated by Just's own special tragedy and frustration: no one with power in America would take him seriously for what he desired to be, and did so well. Liberal supporters consistently tried to remake him as a "model for his race," a man who would selflessly and willingly suppress or abandon his research to render service by teaching humble and grateful blacks headed for the practical work of medicine. Manning notes the inconsistency, even the hypocrisy:

> The foundations were all too ready to suspect blacks of being elitist, separationist, disloyal to their people, but no such doubts arose when it came to supporting whites. Invariably, more delicate questions were asked about blacks than about whites—and the right answers were expected, indeed demanded. The Rockefeller Foundation gave large grants to the eugenicist Raymond Pearl, but no one expected him to identify with poor whites in Appalachia. Rosenwald supported the work of the philosopher Morris Cohen, but he did not inquire about Cohen's ethnic loyalty or suggest that he had to identify with unfortunate Jews living in East Side ghettos. [Rosenwald was also Just's major supporter, and he did ask continuous questions—and demand documentation—about Just's ethnic loyalty.]

We may gain some insight in miniature into Just's special burdens, and the frustration that eventually drove him to European self-exile, by considering his relationship with three key figures, all well documented by Manning: the scientists Jacques Loeb and F.R. Lillie, and the leader of medical philanthropy and reform, Abraham Flexner.

Loeb should have been, and once was, Just's most zealous supporter. This great European biologist and emigré to America was as firm and active in his radical politics as in his research—and Just was America's best biological model for his egalitarian views. Just met the powerful Loeb in 1912 and secured his enthusiastic support. Calling Just "certainly a superior man," he politicked, in his dealings with Flexner among others, for substantial financing both for Just's research and, through Just, for Howard's medical program. Loeb was also active in the NAACP, and was instrumental in securing for Just the unexpected award (for other blacks were far better known) of the first Spingarn Medal in 1915, still the association's highest honor.

Yet by 1923 Loeb had turned so viciously against Just that he scotched the one reasonable chance that Just ever had to secure a major research post at a white establishment, the Rockefeller Institute (now University) in New York. Asked for his candid opinion about Just's suitability, Loeb said that although he had once tried to "help and encourage" Just, he had now concluded "that the man is limited in intelligence, ignorant, incompetent, and conceited; in fact, his so-called research work is not only bad but a nuisance." Loeb suggested that Just become a high school science teacher. Manning comments:

> If there ever had been a chance of Just getting a position at the Rockefeller, Loeb squelched it. Such an appointment would have been a first—symbolic for the whole black race. A different letter could have changed the course of Just's career, and no doubt would have affected the role of blacks in American science generally.

Such vagaries make history and constitute its fascination as an "iffy" science.

What had happened between 1915 and 1923? One might be tempted to charge Loeb with hypocrisy or inconsistency, but the only proper accusation is myopia. I know Loeb's type so well; I have seen men like him so many times during my own career in science. Quite simply, Loeb was the great reductionist of American embryology and physiology and Just, who was developing a contrary holistic view of biological organization, had come to oppose Loeb's biological work. He had even dared to confront Loeb with scathing and convincing counterexamples at public scientific meetings. Higher principles and general political commitments be damned; the man who dares to attack my research is anathema. I know the type so well.

Abraham Flexner was the guru of American medical education during all the years that Just sought financial support for his work and for research in biology at Howard. Flexner wrote the famous 1910 report that changed the face of medical education in America, and later became the first director of the Institute for Advanced Study in Princeton, where he was responsible, in large part, for bringing Einstein to America. Flexner was ubiquitous. He seemed to head, or to dominate through simple membership, the board of directors for every philanthropy and fund that might give money to an American black biologist. His crushing paternalism toward blacks soon turned him against a man of Just's complicated temperament, and he personally blocked support for Just on many occasions, fearing that Just's initial successes in securing support had emboldened him to a point of nonobsequiousness inappropriate for blacks. We may be "going a little too fast," Flexner warned; for there is "a little more danger of spoiling an exceptional colored worker than . . . of spoiling an exceptional white worker."

The story of Just's relationship with F.R. Lillie is the saddest and most revealing of all; for if Just had any true friend among powerful whites of American science and philanthropy, Lillie was the man. Lillie stood by Just throughout his life, always defending him before funding tribunals and research councils, even after the dramatic incident of Just's final, angry departure from Woods Hole—where, on June 27, 1930, at a celebration to honor Lillie's sixtieth birthday and his long directorship of the laboratory, Just gave a prepared speech about his research, then stepped off the podium and announced extempore, apparently still under the spell of six previous euphoric months in Europe after his increasing difficulties at Woods Hole: "I have received more in the way of fraternity and assistance in my one year at the Kaiser-Wilhelm-Institut than in all my other years at Woods Hole put together." The next day, Just left Woods Hole without saying goodbye to anyone, not even to Lillie. He never returned. Yet Lillie remained loyal, and exerted great effort to repatriate Just after the Nazi takeover of France.

Even so, Lillie retained his paternalistic attitude and never understood, until it was too late, Just's passion and need for research (how can any scientist, himself infected with this drive, as Lillie surely was and all of us worth anything are, possibly not understand?). He continually urged Just to pay more attention to teaching at Howard and less to research at Woods Hole. Lillie was the one man who could have solved all Just's institutional problems at a stroke by creating a permanent position for him at Woods Hole as assistant director. He seriously considered this course, but as a cautious and conservative man, finally decided that a black scientist might not be sufficiently acceptable, and that, in any case, Just would be better off "serving his people" at Howard.

Yet, by the late 1930s, prompted perhaps by the dramatic event of 1930, Lillie had finally understood—though his recognition came far too late, psychologically, for Just. In 1938, and in response to one of Just's last attempts for American funding, Lillie wrote in support:

> Just has qualities of genius; nothing whatever turns him aside from his purpose. I have attempted over and over again to get him to conform more to the conditions which his race and the nature of university life in America impose. I think now that this attempt was unwise; certainly it was futile.

And he wrote in Just's obituary notice:

> An element of tragedy ran through all Just's scientific career due to the limitations imposed by being a Negro in America, to which he could make no lasting psychological adjustment in spite of earnest efforts on his part. The numerous grants for research did not compensate for failure to receive an appointment in one of the large universities or research institutes. He felt this as a social stigma, and hence unjust to a scientist of his recognized standing. In Europe he was received with universal kindness, and made to feel at home in every way; he did not experience social discrimination on account of his race, and this contributed greatly to his happiness there. Hence, in part at least, his prolonged self-imposed exile on many occasions. That a man of his ability, scientific devotion, and of such strong personal loyalties as he gave and received, should have been warped in the land of his birth must remain a matter for regret.

Just was buffeted about all his life, often without his knowledge, by men like Loeb, Flexner, and Lillie. Yet his remarkable will and assiduous gift of self-promotion helped him to survive in America and kept him active in research, until Europe beckoned convincingly, and his research, if anything, flourished more.

Yet, as Manning shows so well, we must avoid the temptation to read the story of Just's life as a morality play with Just as a thoroughly exemplary, long-suffering, blameless hero crushed by forces beyond his control. Many of these themes are central and controlling, particularly the pervasive paternalistic racism that Just could not hope to overcome in America. But Just was also an ambiguous and contradictory man—and by no means an entirely sympathetic character. He was deeply conservative in temperament and outlook, and never understood the politics of racism (his first published schoolboy essay extolled the benefits of monopoly in American economic life). Thus, he never sought solidarity in action with other American blacks, but simply persevered in his struggle to support his own research. When blocked or treated unfairly, he would often lash out in self-destructive anger, as at Woods Hole in 1930, never seeming to grasp his larger situation.

Just demonstrated an almost incredible naiveté, blinded no doubt by the understandable joy he felt for his new life in central Europe, towards the growth of fascism in Germany and Italy. He actively sought and won the friendship of the former Hohenzollern crown prince, the son of Kaiser Wilhelm II, and participated in two monarchist gatherings at his home, one with strong Nazi undertones. He approached Mussolini directly with an appeal for funding, praising Il Duce in an embarrassing and obsequious tone. He showed more than a small streak of anti-Semitism, especially when seeking funds from sources known for their own antipathy toward Jews.

Yet this complexity makes Just's story all the more interesting and instructive. We modern liberals, perhaps bathed in our own hypocrisy, might feel more sympathetic toward a fighter like Jackie Robinson than a man like Roy Campanella who just wanted to play ball as best he could and was willing to suffer the indignities of racism in silence—as we might prefer a King to a Just. But didn't Campanella have as much right as Robinson to his place with the Dodgers and his niche in the Hall of Fame—for he was every bit as good a ballplayer as Robinson, and in a world of true justice surely nothing else would count.

Just also deserved his birthright—the chance to compete and triumph by virtue of his clear genius—whatever his dubious personal politics. The last twenty-five years of the civil rights movement embodies a telling irony: a few people like Just, blacks comparatively well off and with a good education, have won a world of opportunity. A black valedictorian at Dartmouth can now write his own ticket; Just was permanently stymied from the start. He didn't seem to care much for the plight of poor blacks, the group that has won so little from our contemporary struggle. Yet he is the kind of man who would have benefited most. The painful story of Just's life underscores the most important question of all: how can we secure these benefits for everybody?

12

Exultation and Explanation

In his most famous article, G.E. Hutchinson, unquestionably the world's greatest living ecologist, emphasizes the fundamental theme of his science by citing an anecdote of J.B.S. Haldane. The great British biologist 'found himself in the company of a group of theologians. On being asked what one could conclude as to the nature of the Creator from a study of his creation, Haldane is said to have answered, 'An inordinate fondness for beetles.' "[1]

Since Linnaeus set the modern style of formal naming in 1758, more than a million species of plants and animals have received Latin binomials. More than 80 percent of these names apply to animals; of the animals, nearly 75 percent are insects; of the insects, about 60 percent are beetles. I am continually amazed that a single mode of organic architecture should engender such diversity. From the tiny trichopterygids less than 1/100 inch in length, to the all-female *Micromalthus* reproducing as a larva in rotting wood, to the quarter pound Goliath beetle and nearly foot long *Batocera* of New Guinea, beetles embody nature's finest display of her principal theme—multifarious diversity. The science of ecology probes

A review of *The Kindly Fruits of the Earth: Recollections of an Embryo Ecologist* by G. Evelyn Hutchinson, and *An Introduction to Population Ecology* by G. Evelyn Hutchinson.

[1]"Homage to Santa Rosalia, or why are there so many kinds of animals," *American Naturalist,* volume 93 (1959), pp. 145–59.

this richness for regularities. As Hutchinson asked in his subtitle to the article graced by Haldane's anecdote: "*Why* are there so many kinds of animals?"

Ecologists must live in tension between two approaches to the diversity of life. On the one hand, they are tempted to bask in the irreducibility and glory of it all—exult and record. But, on the other, they acknowledge that science is a search for repeated pattern. Laws and regularities underlie the display. Why are there more species in tropical than in temperate zones? Why so many more small animals than large? Why do food chains tend to be longer in the sea than on land? Why are reefs so diversely and sea shores so sparsely populated with species? As an explanatory science, ecology traffics in differential equations, complex statistics, mathematical modeling, and computer simulation. I haven't seen a picture of an animal in the leading journal of evolutionary ecology for years.

Many ecologists have escaped this tension by focusing their work on a single approach—exultation or explanation—and by treating the other side with territorial suspicion and derogation. Hutchinson has practiced and loved both all his life. He had all the advantages of an upper-class, English, intellectual background. *The Kindly Fruits of the Earth,* an autobiographical account of his early life up to his appointment at Yale in the late 1920s, reads like an extended idyll. More than half the book traces his education through the English public schools to Emmanuel College, Cambridge (his father had been a mineralogist and don at Pembroke College). The last three chapters recount his early professional experiences in Italy and South Africa. Hutchinson apparently enjoyed it all, from botanizing in the woods to tea with a variety of notables.

Perhaps one has to be a bit more dyspeptic or analytic than Hutchinson to communicate the meaning of such advantages to those of us who did not grow up enjoying them. Moreover, since Hutchinson is as legendary for his kindness as for his brilliant mind, this memoir only records the good deeds of dead people—for if he follows the old motto, *de mortuis nihil nisi bonum* (say only good of the dead), of the living he chooses to say absolutely nothing at all! (Hence the book must exclude Hutchinson's last half century—the period of his greatest achievements.) This benign style may not win many readers beyond the large group of professionals who admire Hutchinson and who wish to learn more about the illustrious scientists associated with him, but I am somehow comforted to discover that such elemental decency actually exists.

During his early years at Yale, Hutchinson worked primarily in biogeochemistry, insect classification, and limnology.[2] Ecology, when Hutchinson entered the field, was largely a profession of empirical recorders. It boasted few organizing concepts beyond "succession theory"—a somewhat mystical idea that treated communities of species as "superorganisms," and tried to identify an orderly progression of communities in a history of colonization within a geographic region. The foundations of modern population ecology had been established in equations of demography and population growth by such mathematicians as A.J. Lotka, V. Volterra, and G.F. Gause, but most ecologists either didn't know the work or questioned its relevance to real field data.

Hutchinson had a major part in fostering the two major directions of modern theoretical ecology. One approach, production or systems ecology, emphasizes the flow of energy among the components of ecosystems, for example the efficiency of "conversion" between levels in a food chain—how many calories of green plants are needed to sustain herbivores on land, what weight of consumed herbivores maintains a population of carnivores, etc. Raymond Lindeman, the Pergolesi of ecology, wrote the founding document of this discipline, but died in 1942 at age twenty-six while it was still in press.[3] He developed many of his ideas as a post-doctoral student with Hutchinson. Moreover, after the journal *Ecology* had rejected Lindeman's paper on the advice of two traditional ecologists, Hutchinson's intercession proved crucial to a fa-

[2]I thank my colleague Robert E. Cook for sharing his insights with me about the history of twentieth-century ecology and Hutchinson's role in it.

[3]R. Lindeman, "The trophic-dynamic aspect in ecology," *Ecology*, vol. 23 (1942), p. 399.

vorable reconsideration. Hutchinson wrote a private letter to the editor of *Ecology*[4] that remains a stirring defense of theoretical work as a guide for empirical research, not a denial of its importance. He ends with Sir Thomas Browne's defense of fruitful error—"the certainty hereof let the arithmetick of the last day determine."

Hutchinson played a more direct role in establishing the second approach, population ecology. This discipline begins with the central Darwinian postulate that nature manifests no higher principle than the struggle of individual organisms to maximize their own reproductive success. Notions of community and natural harmony, however illuminating as metaphors, do not reflect nature's primary evolutionary unit, the population of individuals within a species. Modern ecology begins with equations of population growth, including differing rates of fertility and mortality. It attempts to build toward a theory that will explain the size of populations, their rates of change in size, and the co-existence of many populations in a single area through competition and avoidance. *An Introduction to Population Ecology* presents Hutchinson's recent survey of this central subdiscipline in modern evolutionary theory. It is a textbook, and lay readers (and some professionals as well) may have to review their algebra before tackling it. But, unlike most examples of the genre, this book is accessible, intelligent, and enjoyable to read as literature. As a word, ecology has been so debased by recent political usage that many people employ it to identify anything good that happens far from cities and without human interference. But among professionals ecology is the science that tries to understand why there are so many kinds of organisms and why some habitats and some parts of the world harbor more kinds than others. It attempts, in other words, to explain the spatial and temporal structure of organic diversity. Its fundamental concept is the *niche*—an expression of the location and function of a species in a habitat.

Hutchinson made his most important contribution to theoretical ecology by establishing a workable, quantifiable con-

[4]Published in R.E. Cook, "Raymond Lindeman and the trophic-dynamic concept in ecology," *Science*, vol. 198 (1977), pp. 22–26.

cept of niche.[5] Before him, ecologists debated whether a niche represented an organism's address or its profession—in other words, are niches environmental spaces that exist whether or not organisms live in them, or are they created by the range of activities (feeding, nesting, etc.) performed by each species in its unique way? If niches are addresses, then the problem of organic diversity might be reduced to a study of physical habitats and their change through space and time. If niches are professions, then diversity will be set, in large part, by the kinds of organisms that settle in an area. We must then wonder whether a nearly infinite subdivision of niches might be possible and whether the notion of a limit to diversity makes any sense at all.

Not surprisingly, Hutchinson's definition embodied fruitful aspects of both views—organisms do create ecospace through their activities, but the nature of physical space and resources sets important limits. Hutchinson measured the niche of a species by defining graded components in the environment that must be important for survival—temperature, particle size of food, or nesting height in trees among birds, for example. He then depicted these environmental components as mathematical axes at right angles to each other. (Axes at right angles are mathematically independent and, although we lose the pleasure of visualization, an axis may stand at right angles to as many other axes as we like in appropriate spaces higher than three dimensions.) For each axis, we can plot the total range used by a species; the volume of space defined by the intersection of these ranges for each axis is the niche. Since useful quantification is so often the key to fruitful science, this concept of niche as a definable segment of resource space has become the basis for studies of diversity and its limits.

After developing this definition, Hutchinson noticed[6] that closely related species inhabiting the same area often space themselves out quite regularly along individual resource axes. This led to what I regard as the most arresting and important

[5]Concluding remarks, *Cold Spring Harbor Symposium in Quantitative Biology,* No. 22 (1957), pp. 415–27.
[6]In his seminal article cited in note 1.

concept in our attempt to understand the structure of diversity—limiting similarity. Is there a maximum similarity (or niche overlap) beyond which two related species cannot coexist in the same area? If such limiting similarity can be defined, then we will have a theory of maximum organic diversity. Ecologists have not yet developed a general theory of limiting similarity. In the absence of a general theory, we might probe empirically for any regularities in the similarity of related species living together in a single area and feeding at the same level of a food web. Again, Hutchinson was a pioneer. He noticed that such related species often differ by a length ratio averaging 1.28—that is, the larger species is 1.28 longer than the smaller for critical dimensions related to feeding. This length ratio corresponds almost exactly to a doubling of weight. (Length varies as the cube root of weight, and the cube root of two—corresponding to a doubling of weight—is 1.26.)

Perhaps Hutchinson has even discovered something fundamental about the structure of similarity in general systems. Princeton ecologists Henry Horn and Robert May have provided a compendium of further biological cases for the 1.26 ratio in biological systems.[7] But they also note that length ratios for consorts or ensembles of similar musical instruments meant to be played together tend to obey the rule—the sequence of recorders, sopranino, descant, treble, tenor, bass, and great bass, runs 1.2, 1.5, 1.3, 1.6, and 1.4. The four stringed instruments of the modern orchestra are way off; they increase too rapidly in length (1.8 for cello to viola and 1.7 for bass fiddle to cello). But these instruments were not designed to play as a group. Carleen Hutchins has built a new "violin family" of "acoustically matched stringed instruments" analogous to the old consorts and meant as an ensemble. The eight instruments have successive ratios of 1.2, 1.2, 1.3, 1.3, 1.3, and 1.3. Horn and May also measured a sequence of 1.3, 1.2, 1.3, and 1.2 in wheels of a tricycle-bicycle sequence and 1.2, 1.2, 1.2, and 1.3 for a single manufacturer's set of iron skillets. Who knows what all this means; I do not think it is an accident.

[7]Henry S. Horn and Robert M. May, "Limits to similarity among coexisting competitors," *Nature*, vol. 270 (1977), pp. 660–61.

Horn and May conclude that Hutchinson's rule "may well derive from generalities about assembling sets of tools, rather than from any biological peculiarities."

This discussion of theory presents only half of Hutchinson. It helps to explain why he is a great scientist, but it does not capture the special character that makes him a great man as well, an object of that rarest of all attitudes among scientists—reverence. For Hutchinson exults as much in the pure diversity of all knowledge as in the hunger to explain nature's patterns. He is, to use an expressive English word virtually unknown in America, a polymath. He cherishes every detail of every subtle difference among objects and casts his web of knowledge over more disciplines than many colleges teach. His favorite themes, to express the subject biologically, are diversity and ornamentation—seeing the great and general in the small and the detailed. He wrote an entire book on ornament in Tibetan culture.[8] For more than ten years he wrote a column for the *American Scientist* entitled "Marginalia," an extended and delightful set of prose footnotes and general wanderings around the periphery of biology and human life. Now he works on literal marginalia—the illuminations of plants and animals that adorn so many medieval manuscripts.

The love of detail for its own sake pervades both books. The textbook on population ecology is adorned with historical footnotes and tangential wanderings on etymology and minutiae of natural history. Some pages even attain the scholarly ideal—all footnote with no text at all! They give to the genre that rarest of traits—a human face—and make this textbook a delight rather than a chore to read.

Hutchinson's emphasis on unevaluated detail is not so successful in his autobiographical memoir—for it *is* the content here, rather than a commentary upon something more coherent. Hutchinson may view all bits of knowledge as equal in interest and impact. More ordinary mortals tend to be choosy. Even I, a New Yorker who rarely missed the San Gennaro

[8] *The Clear Mirror: A Pattern of Life in Goa and in Indian Tibet* (Cambridge University Press, 1963).

festival, and who even mastered the nickel-through-water-into-cup game, flagged a bit after ten pages on whether and, if so, how and why the relic of the blood of St. Januarius liquefies during holy times in Naples. And much as I love the Onycophora, that curious group of potential intermediates between worms and insects, I wished the several pages on their detailed taxonomy were shorter and the paragraph on race relations in South Africa much longer. Still, such excursions are more than amply compensated by the insights liberally scattered throughout this memoir.

Many scientists undoubtedly view this discursive side of Hutchinson as marginal to his "real" accomplishments in constructing theories and explanations. Henry Ford's equation of history and bunk ranks as good mathematics for many professionals who hold the false view that science progresses by gathering objective information and treating data with a timeless device known as the scientific method. They might regard Hutchinson's remarkable brand of scholarship as commendable but surely irrelevant to professional activity.

Such an attitude compromises science, as much as its disheartening Philistinism holds the ideals of learning in contempt. An experimentalist, using technological machinery under a strict and repeated protocol, might get by in ignorance of the history and implications of his field. Those who work directly with nature's multifarious complexity cannot afford such narrowness. We are tied to historical habits of thought (doctrines of progress, gradual change, and linear causality, for example) and methods of procedure (reduction to component parts as a mode of explanation, rather than direct study of interaction). When we fail to recognize that these are habits of inertia rather than nature's truths, new paths are closed off. Scientists ignorant of history are not so much condemned to repeat it, as to be confused and unenterprising. Hutchinson's work has been richer for its combination of explanation and exultation.

In the coda to his textbook, Hutchinson discusses the solitaire, a bird extinguished by European settlers on the island of Rodriguez in the Indian Ocean. According to an accurate early eighteenth-century chronicler, solitaires mated for life and held a territory 200 yards or so in radius. Each pair incubated a single egg and reared the helpless offspring for several months after hatching. Hutchinson continues:

> We now reach the extraordinary part of the story. Leguat [the chronicler] says that when the chick had left the nest for some days, a company of 30 or 40 adults brought another chick and, being joined by the parents, the two young birds were escorted to some unoccupied territory. This extraordinary ritual he describes as the marriage of the solitaire. Leguat seems so reliable on most other aspects of the bird's biology, and the marriage, if it occurred, must have been so conspicuous, that it is hard to doubt its reality.

Shall we regret the solitaire's passing because its unique behavior might have suggested new generalities in the currently "hot" field of adaptive strategies in mating behavior? Or shall we simply mourn the lost opportunity for watching something so fascinating and so different? We had best cherish exultation and explanation with equal tenacity, though I myself would trade several good generalities for the chance to witness such a spectacle.

13

Calling Dr. Thomas

Lewis Thomas's autobiography is, in many good ways, a deceptive book. Its two parts read, superficially, like a reverie, full of cozy nostalgia celebrating a life well lived. The first section describes his father's career as a general practitioner in Flushing, New York City—my old stomping grounds when I was a boy, and well remembered as densely urbanized in the late 1940s, but virtually a rural village during Thomas's youth. We read of horse-drawn carriages, home visits, endless concern, and little pecuniary reward. The second section treats Thomas's adult career as a doctor and medical administrator. He transmits an overwhelming impression that research can be fun, and he even manages to intimate that administration can be more rewarding, even more amusing, than frustrating—a proposition that I find hard to believe.

Yet in many ways, and despite its overt content, *The Youngest Science* is a profound and even disturbing book. Both its parts are fundamentally about trade-offs made necessary, and surely in large part desirable, by improving technology—specifically, the sacrifice of heart for efficiency. Dr. Thomas, Sr., and his old medicine could do precious little to cure disease, but his ministrations were often effective because his personal concern and endless attention inspired confidence. In the second part, Thomas describes his own research as especially enjoyable, and perhaps even more effective, because its small scale, and

A review of *The Youngest Science: Notes of a Medicine Watcher* by Lewis Thomas.

LEWIS THOMAS

the limited resources that make a senior researcher scrounge for equipment and do the routine procedures himself, guarantee a closeness to detail that may be all for the good (ethically and factually). By contrast, modern heads of laboratories command empires and may never sit at a workbench. Vastly more work can be done in a given amount of time, but senior researchers can so lose touch with the day-to-day procedures of their laboratories that the fraudulent data of a Darsee can go undetected for years.

Thomas is able to insinuate these serious themes into a

charming and superficially rambling narrative because he has been perfecting this technique for years in books of essays that have made him a popular figure—*The Lives of a Cell* and *The Medusa and the Snail.* Thomas's essays are unsurpassed for their uncanny knack of starting with a simple fact, a bit of home-grown Yankee wisdom, and slyly developing its implications until some profound truth slips through before you hardly notice. Thomas has now successfully parlayed this strategy into book length.

I was enthralled by the book's first part, because I harbor memories of the last years of physicians' home visits in my own version of Flushing. (I distinctly remember how Dr. Schildkraut's arrival to treat me for flu once interrupted my fascination with the radio news of King George VI's death.)

Dr. Thomas, Sr., began with a bicycle, then graduated to a horse and buggy and finally, a year before Lewis's birth, to an automobile. He never had an office nurse or secretary. He kept office hours at home from one to two in the afternoon and from seven to eight in the evening, seeing an average of ten patients each hour. The rest of the day he was on the road, making rounds at the local hospital early in the morning and performing his house calls throughout the rest of the day. At night, long after the rest of his family had gone to bed, he kept his accounts and fretted over the characteristic late (or non) payment of most clients. This routine was interrupted at least once, and often three or four times each night, by emergency calls requiring more house visits. Lewis Thomas remembers how his father would usually swear "Damnation" under his breath, and how he sometimes even got angry enough to exclaim, "God damn it." Yet he always went.

The monetary rewards for all this effort were modest. Thomas's best friend in the Harvard class of 1937 put out the yearbook and sent a questionnaire to Harvard medical graduates from the previous classes of 1927, 1917, and 1907. The average income for the ten-year graduates was $3,500, rising to $7,500 for the twenty-year men. "One man, a urologist, reported an income of $50,000, but he was an anomaly; all the rest made, by the standards of 1937, respectable but very modest sums of money." Medicine, the old saw went at the time, was an honorable profession, but hours were long, monetary reward modest, and effectiveness limited.

Limited effectiveness may be the most striking difference between medicine then and now. This theme also provides the source for Thomas's title, *The Youngest Science,* since it was only with the introduction of sulfa drugs in the late 1930s and penicillin and other antibiotics a few years later that doctors finally learned to treat and cure more than a very few diseases.

Old Dr. Thomas, for all his efforts and hours, could do very little to help his patients directly. Lewis Thomas describes his childhood conversations with his father, who often took him along on house calls:

> The general drift of his conversation was intended to make clear to me, early on, the aspect of medicine that troubled him most through his professional life; there were so many people needing help, and so little that he could do for any of them. It was necessary for him to be available, and to make all these calls at their homes, but I was not to have the idea that he could do anything much to change the course of their illnesses.

In those days, a doctor's reputation as healer could be assured because we all slip so easily into the *post hoc, ergo propter hoc* fallacy. Most diseases run a natural course and few lead to inevitable death. Thus large numbers of people will get well, no matter what ministrations doctors provide. But the belief that a doctor's attention makes a difference—especially back in the old days when they came to your house, knew your dog's name, and spent endless time in encouraging discussion—can be powerful, especially when that attention is followed by a natural recovery. This situation persisted right through Lewis Thomas's time in medical school. He writes of medicine just before the introduction of sulfanilamide:

> It gradually dawned on us that we didn't know much that was really useful, that we could do nothing to change the

course of the great majority of the diseases we were so busy analyzing, that medicine, for all its facade as a learned profession, was in real life a profoundly ignorant occupation.

With the exception of morphine and digitalis, most drugs in his father's arsenal were placebos, but none the less effective for that. The "placebo effect" is one of the most interesting phenomena in medicine. Some placebos "work" because people are getting better anyway, but states of mind inspired by confidence clearly influence the course of health in other cases. Placebos sometimes really make a difference, if we have confidence in the doctor—hence the secret of effectiveness for the old-fashioned approach designed to inspire such faith. Thomas writes of the mystery and ritual involved in writing prescriptions, invariably in illegible Latin:

> They were placebos, and they had been the principal mainstay of medicine, the sole technology, for so long a time—millennia—that they had the incantatory power of religious ritual. My father had little faith in the effectiveness of any of them, but he used them daily in his practice. . . . They did no harm, so far as he could see; if nothing else, they gave the patient something to do while the illness, whatever, was working its way through its appointed course.

The principal goal of practical medicine therefore became accurate diagnosis, rather than treatment or cure. If medicine could do little to alter the course of most diseases, at least a doctor could tell a patient what to expect, and explain his chances and probable progress with accuracy and compassion. Thomas writes of his own training: "Our task for the future was to be diagnosis and explanation. Explanation was the real business of medicine. What the ill patient and his family wanted most was to know the name of the illness, and then, if possible, what had caused it, and finally, most important of all, how it was likely to turn out." Of a master doctor and diagnostician, he writes: "So far as I know, from that three months of close contact with Blumgart for three hours every morning, he was never wrong, not once. But I can recall only three or four patients for whom the diagnosis resulted in the possibility of doing something to change the course of the illness."

Medical men did, of course, try a variety of methods to alleviate disease. But most were useless fads, and Thomas tells some amusing stories. In the first decade of this century, the notion arose that most human disease might be attributed to the absorption of toxins from the lower intestinal tract. This "autointoxication" could be countered by keeping the large bowel empty, and a variety of devices were introduced to assure such regularity. In 1912, Thomas's father bought a curious item from a medical supply house—a large lead object the size of a bowling ball, wrapped in leather. The patient would lie flat in bed and, several times a day, roll the ball clockwise following the course of the large intestine. Needless to say, it didn't work, and Thomas, Sr., eventually tossed it out. Twelve years later, a local newspaper announced the discovery of a cannonball from the Revolutionary War, unearthed in the garden behind the Thomases' neighbor's yard. No one could figure out how the lead sphere got there because it followed no known trajectory of any engagement from any war.

As a consequence of personal prejudice, I found the second part of *The Youngest Science,* Thomas's description of his own professional life, less interesting than the first. Thomas has done some important research, primarily on immunological questions, but he has spent most of his life as a medical administrator, in a long and impressive series of deanships and directorships of such institutions as NYU, Yale, and the Memorial Sloan-Kettering Cancer Center. Now I know that any institution would fall apart without its officers (the key person, as Thomas would readily admit, being the chief administrative secretary, not the boss), but I've never been able to develop much interest in the process, however indispensable it must be. In part, Thomas himself identifies the source of my malaise: a well-oiled place is best served when left alone.

The governance of academic institutions has been considered and reconsidered, reviewed over and over by faculty committee after committee, had more reports written about it than even the curriculum, even tenure. Nothing much ever comes of the labor. How should a university be run? Who is really in charge, holding the power? The proper answer is, of course, nobody.

If real answers existed, someone would have figured them out by now, since universities are usually run by people of reasonable intelligence and good will. The only exit from this paradox must be an admission that no good, or unambiguous, solutions exist. And you can only write so much about unresolvable questions.

I enjoyed the chapters on Thomas's research even more—especially because he discusses his failures as much as his successes and continually emphasizes the pure fun of research. As any practitioner knows, the effort of research is so tedious and time consuming that when the work stops being fun, there's no sense in continuing. The best scientists live a life of keen amusement.

Underlying all this pleasant discussion is the persistently serious theme of crisis, or at least lamentable loss, in the trade of personal attention for high technology in the practice of medicine, particularly at large hospitals. Nurses, for whom Thomas has the deepest respect and affection, have taken on the task of compassion and personal care that once extended through all medical personnel. In modern hospitals, he notes sardonically, you get greatest individual attention while discussing your insurance coverage in the admitting office but never again on the wards. The awesome and effective machinery of modern medicine seems to place an unsurmountable barrier between doctor and patient. Old Dr. Thomas would stand in amazement before the advances of modern medicine, but would simply not comprehend the loss of personal contact.

Can anything be done to reinsinuate this sense of active compassion into medical care at all levels? Such an accomplishment would be more than merely pleasant; it would be positively therapeutic, since states of mind can influence the course of nearly any disease (as Thomas with his expertise in immunology and the body's own defenses continually emphasizes). Thomas acknowledges, of course—and this is the ultimate irony—that faced with the choice between his dad's knowledge of your dog's name and the most heartless mechanism of a modern hospital, any patient would choose the hospital—for getting well remains the primary desideratum in seeking medical care. But is there no way to have both?

Probably not as a statistical phenomenon, for institutions and machinery create personalities and impose the fragmented and limited contact that most medical people now have with their patients. But if medicine attracted more rational humanists, then a force of strong personal commitment might just balance or even overcome the shaping role of institutions. Thomas himself is the prototype for such a man, a medical administrator (stereotypical role for the heartless) who exudes in his writing and exemplifies by his life the old-fashioned traits of care and compassion that we must try again to bring into prominence. Perhaps we can only promote this rediscovery by example. If so, we will find no better case study than this book. Thus, though Thomas did not write *The Youngest Science* to toot his horn—for he seems the very embodiment of a genuine modesty that does not deny skill and accomplishment, yet refuses to use these blessings for one-upmanship—its ultimate message loudly proclaims that what medicine needs most is an army of Lewis Thomas clones.

V

In Praise of Reason

14

Pleasant Dreams

A recent cartoon shows two aged scientists sharing a pipe in the smug satisfaction of a life well lived. "One thing I'll say for us," exclaims the first, "we never stooped to popularizing science." I don't deny that I have known such men, but their number is far smaller than most people think. Most scientists do wish to transmit their information and their excitement to nonprofessionals. If few actually write or speak for the public, their reticence arises more from shyness and inexperience than from lack of concern.

The great works of popular science are lucid expositions of difficult subjects in nontechnical language—Bertrand Russell's *The ABC of Relativity,* or George Gamow's series about the adventures of Mr. Tomkins in a world where the physics of relativity and quanta rule over objects at human scale. These works clarify the content of science, but do not make the process of scientific discovery any less mysterious. Science might, after all, produce clear messages by using arcane procedures accessible only to an initiated priesthood. To break down this final barrier between science and its public, scientists must present themselves as well as their work. And here, at the threshold of autobiography, most scientists balk. They may produce *in camera* works full of unconscious distortion (as Darwin did in writing, for his children, an autobiographical

A review of *Disturbing the Universe* by Freeman Dyson.

FREEMAN DYSON

note never intended for popular consumption). Or they may discourse in wooden, unrevealing words about their fascinating lives (L.S.B. Leakey's *By the Evidence,* for example), or write only to vindicate their positions in a lifetime of petty squabbles. Usually, of course, they write nothing, repeating the mute response of Jesus to Pilate's question, "What is truth?" rather than Pilate's statement upon displaying Jesus after torture—*ecce homo,* behold the man. Yet science, as an activity, will remain inaccessible so long as scientists refuse to speak honestly about their own lives and dreams.

Freeman Dyson has broken a path by showing that the composition of a candid autobiography can be fun, or at least cathartic. Dyson, distinguished physicist and professor at the Institute for Advanced Study in Princeton, New Jersey, has lived by the motto he has chosen for our potential salvation—diversity. He has divided his scientific time between theoretical physics and practical (or impractical) applications. He has tried to design safe nuclear reactors, championed an indefensible scheme to propel spacecraft with atomic explosions (bombs with all their attendant fallout). He played a dubious part as something of a military hard liner (now reconstructed) during the cold war. He has ranged in his public positions from a stalwart defender of open-air atomic testing to the favorite physicist of visionaries who would colonize the solar system (if not the universe), using what Dyson calls "green" rather than "gray" technology.

Dyson, son of a composer and a lawyer, attended the prestigious public school of Winchester. (He writes as if such advantages are a natural concomitant of growing up.) He was too young, and in the wrong place, to join his colleagues in building nuclear bombs—and thus "knowing sin," as Oppenheimer said and he spent the war working with statistics of death and destruction for the British Bomber Command. But he came to America after the war, and met Oppenheimer, Teller, Bethe, and Feynman not at the moment of their legitimate triumphs, but in their most trying and often, alas, their worst hours, during the cold war. This perspective, of someone who came just after, someone who never quite hit the stratosphere, strikes me as the most interesting feature of Dyson's book. The heroes have more choices. At the lower level of the young and the struggling, science as an institution is more clearly revealed.

By combining this perspective from below with unflinching honesty in recollection, Dyson's book serves as a chilling testimony to the forces that lead so many decent men to compromise and slip deeper and deeper after a poorly rationalized first plunge. Dyson, though a pacifist by belief at the beginning of World War II, joined the Bomber Command as a young man who, but for the grace of God and social class, might have been inside the planes rather than calculating their chances of returning from the Battle of Berlin. He quickly recognized three sobering facts: that the missions were ineffective, that experience brought no higher rate of survival per mission (and that probability of death was therefore a grim and inexorable function of missions flown), and that such simple changes as reducing the crews and dismantling some heavy equipment would save many lives.

He made some mildly protesting noises about these facts to his boss, but retreated to frustrated obedience when military bureaucracy stifled any chance for reform. But a colleague who recognized that lives might be saved simply by enlarging the bombers' escape hatches fought for two years and prevailed even though the war had nearly ended. In a passage that impressed me as a powerful, personal epitome of the central dilemma faced by good intention in a world where brutality abounds and self-sacrifice (even in small things) is unfortunately as rare as rationalization is easy, Dyson writes:

> I began to look backward and to ask myself how it happened that I let myself become involved in this crazy game of murder. Since the beginning of the war I had been retreating step by step from one moral position to another, until at the end I had no moral position at all. At the beginning of the war I . . . was morally opposed to all violence. After a year of war I retreated and said, Unfortunately nonviolent resistance against Hitler is impracticable, but I am still morally opposed to bombing. A few years later I said, Unfortunately it seems that bombing is necessary in order to win the war, and so I am willing to go to work for Bomber Command, but I am still morally opposed to bombing cities indiscriminately. After I arrived at Bomber Command I said, Unfortunately it turns out that we are after all bombing cities indiscriminately, but this is morally justified as it is helping to win the war. A year later I said, Unfortunately it seems that our bombing is not really helping to win the war, but at least I am morally justified in working to save

the lives of the bomber crews. In the last spring of the war I could no longer find any excuses.

The theme of apology continues as Dyson mounts the ladder of fame and influence. He undertook a public campaign against a nuclear test ban treaty—a position he now regards as "wrong technically, wrong militarily, wrong politically, and wrong morally"—both as an act of personal loyalty toward his friend Edward Teller and as a last-ditch effort to save from extinction his pet project for space travel by bomb propulsion. As for Teller, Dyson excuses his decision to testify against Oppenheimer: "A careful reading of his testimony at the trial shows that he intended no personal betrayal. He wanted only to destroy Oppenheimer's political power, not to damage Oppenheimer personally." (But can the two be separated?) Dyson was angry with Teller for a time, but forgave him or so he states, when he heard Teller playing the Bach Prelude in E-flat minor, a piece that Dyson had loved as a child. If only aesthetic and moral sensitivity went hand in hand, I would be able to make much more sense out of this crazy world. I am baffled by Dyson's attempts to justify both Teller and himself, but how can someone judge when he did not live through those times? I was a boy then, and Joe DiMaggio was my hero. And besides, no one has ever asked a paleontologist to construct instruments of terror. I know what I hope I would have done in Dyson's position; I cannot be sure what I would actually have done.

In the last quarter of his book, Dyson presents a series of essays on possible improvements for now and hopes for the future. As an unabashed dreamer, he envisions a possible colonization of asteroids, the preservation of earth as a quaint, ecological abode, and a rapid expansion of colonies throughout the galaxy, if not beyond—and all with a fastidiousness about ecological consequences that would raise no eyebrows upon any editor of the *Whole Earth Catalog.* I can't follow Dyson all the way, or even much beyond the gravitational pull of *terra firma,* but I appreciate both the vision and the biological style of thinking that accompanies these essays.

In the parochial world of academe, characteristic attitudes often define professions (though professions might argue that they convey truth, rather than partial visions shaped by their material). Evolutionary biologists (I am one) tend to equate goodness with what we view as the agent and the result of evolutionary change: the correlation between unconstrained smallness and innovation (for new species usually arise in tiny populations separated from larger parental groups), and the sheer exuberant diversity of life. If an evolutionist believes in any *summum bonum,* it can only be diversity itself (not the attainment of "higher" states, since "progress" of this sort plays no part in modern evolutionary theory). How could I think otherwise as I sit here, with millions of *E. coli* metabolizing in my gut, billions of neurons cogitating in my brain, a cockroach and a centipede on the floor, and a skunk in the garbage outside (yes, even in Cambridge, MA). In our tribalism, we often think that physicists make different equations between the world outside and the true and the beautiful—perhaps because they are more interested in overarching, simplified, and unifying law than in all the messy diversity of a world so largely regulated by the chancy interventions of history.

Perhaps we are wrong about physicists in general, but at least I welcome Dyson into our tribe. He not only believes in smallness and diversity for its own sake; he has also defined his scientific ethic by fighting bureaucracy and institutionalized "big" science as the agents of stultified mediocrity. Dyson is not being facetious when he argues that we have no safe nuclear reactor today because "nobody any longer has any fun building reactors." He worked with a small group of unconstrained enthusiasts for nuclear power in the early days. But the technocrats took over, plans rigidified, economics intervened, and "innovation" now centers upon minor modifications upon a set design.

Yet just as a whale bears vestiges of leg bones, I detect a residual attitude that trips Dyson at the threshold of grasping the biological way of thought. Dyson is filled with a vision of the "rightness" of things. The universe must be ordered. All must exist for a reason and purpose. Everything fits with everything else. But why should such harmony prevail? Does a land snail that is blown by a hurricane to a distant island, fertilizes itself, and becomes the progenitor of a new species fit in some predictable way into an overarching order of things? I don't know what to say about such an event except that it just happened. Nature produces some order by rejecting the ill-adapted, but we can hardly hope to specify an optimal arrangement of adapted species. At this level, life is intrinsically unpredictable; what else does the theme of diversity proclaim?

The whale's leg bones record its terrestrial past, but do not hamper its current function in any serious way. Dyson's vision of rightness may reflect his background in physics, but it does seriously compromise his view of biology in two important senses. First, in his Panglossian world everything that exists must be for the best. In a curious reversal of causality, he argues that we have linguistic variety in order to separate human groups and preserve the ideal of diversity:

> It is not just an inconvenient historical accident that we have a variety of languages. It was nature's way to make it possible for us to evolve rapidly. . . . To keep a small community genetically isolated and to enable it to evolve new social institutions, it was vitally important that the members of the community should be quickly separated from their neighbors by barriers of language.

Yet unless biologists are totally cockeyed, the reverse must be true. Human groups became separated and, in their isolation, diverged in their speech. This divergence may keep groups separate if they come into contact later on, but it cannot be the cause of separation itself, only a result. The current utility of a structure or institution does not specify the reason for its origin. Brains did not evolve so that Bach might write the Prelude in E-flat minor, but Dyson isn't the only one who rejoices that brains can sometimes produce such a masterpiece.

This reversal of causality in Dyson's thinking is a common fallacy of the vision of "rightness" that Voltaire satirized so unmercifully in *Candide* (subtitled *Optimism*). As the foolish Dr.

Pangloss says: "Things cannot be other than they are. . . . Everything is made for the best purpose. Our noses were made to carry spectacles, so we have spectacles. Legs were clearly intended for breeches, and we wear them."

Secondly, Dyson applies his reinstituted doctrine of final causes to an argument for the prevalence of mind in the universe. He argues that mind necessarily enters "into our awareness of nature" at the "highest" level of human consciousness and at the "lowest" level of single atoms and electrons where, by quantum theory, we cannot formulate a description independent of our observation. If mind must be considered at the extremes, then why not in the middle, and everywhere. And, if everywhere, then is not the "rightness" of things a record of the omnipresence of mind? And is not this "rightness" reflected primarily in how well we fit into a universe that existed long before us? "The more I examine the universe and study the details of its architecture, the more evidence I find that the universe in some sense must have known that we were coming."

Dyson then shows how we could not have fit into a different world. If the strengths of physical forces were ordered differently, stars would not burn and life dependent upon solar warmth could not exist. If no "exclusion principle" reigned, and if two electrons could occupy the same state, then "none of our essential chemistry would survive," and life as we know it would fail. But again, our modern Dr. Pangloss has his causality reversed. The universe was here for whatever reason (if any) and we fit in much later. It seems the height of antiquated hubris to claim that the universe carried on as it did for billions of years in order to form a comfortable abode for us.

Chance and historical contingency give the world of life most of its glory and fascination. I sit here happy to be alive and sure that some reason must exist for "why me?" Or the earth might have been totally covered with water, and an octopus might now be telling its children why the eight-legged God of all things had made such a perfect world for cephalopods. Sure we fit. We wouldn't be here if we didn't. But the world wasn't made for us and it will endure without us.

All dreamers must fail; but without the dreamers, I suspect that our earth would not be reverberating with the question "why," simply because its brainiest mammal would be sitting in a tropical tree asking more limited questions about adequate food and shelter for apes. In the *Essay on Man,* not exactly a document for dreamers, Pope writes:

> Oh sons of earth! attempt ye still to
>   rise,
> By mountains piled on mountains
>   to the skies?
> Heaven still with laughter the vain
>   toil surveys,
> And buries madmen in the heaps
>   they raise.

But Pope's madmen are also guardians of the dream. Perhaps the asteroids Dyson hopes to colonize are miserable, useless hunks of rock. Perhaps we will exterminate ourselves before we ever get there. But Lord help us if we lose interest.

15

The Perils of Hope

I spent an hour on Fifth Avenue last week, just a visitor from Boston now, but awed, once again and as always, by the size and vitality of my native city. In the shadow of St. Patrick's, I stood transfixed before the window displays of commercial spinoffs from computer technology—watches that play baseball and beep "Dixie," radios thinner than my bankbook, $10 calculators representing a thousandfold advance upon the $400 device I bought with such a sense of modernity just ten years ago. It took this scale of densely packed, beeping, flashing, almost living and pulsating objects to force my reluctant paleontologist's soul to a recognition that the revolution is already upon us—the most profound change in human life since everything from trains to television brought us all together. One block west, at Rockefeller Center, an inscription proclaims: "Wisdom and knowledge shall be the stability of thy times." I wonder. Wisdom perhaps, but. . . .

Robert Jastrow's *The Enchanted Loom* treats this revolution provocatively and with eloquence. He argues that computers will soon be sufficiently miniaturized and refined to become a genuine extension and improvement upon true human intelligence. An old chestnut proclaims that machines can never equal or even be, in any meaningful sense, at all like the human brain because improvements in computers only add circuitry,

A review of *The Enchanted Loom: Mind in the Universe* by Robert Jastrow.

while organic intelligence is an ineffable, qualitative something that cannot, in principle, be matched by mere quantitative addition. This may be so, but I join Jastrow in defending the plausibility of an opposite interpretation—what we call wit, wisdom, brilliance, and insight need not have, as its material substrate, any more than a vast increase in the number and connectivity of circuits.

Hegelians and Marxists have long advocated the "transformation of quantity into quality" as a basic statement about the nature of change. Graded inputs need not simply yield graded outputs. Instead, systems often resist change and absorb stresses to a breaking point, beyond which an additional small input may trigger a profound change of state. Water at 50 degrees Centigrade is not half boiling. A computer twice as big as another may not simply keep accounts twice as fast. Our metaphor about straws and camels' backs reflects an implicit understanding that not all change is continuous.

The previous impediment, Jastrow argues, was not a technological inability to mimic the brain's operation, but a limitation of size. If organic brains reach human capacities primarily by increasing the number and connectivity of neurons, then computers with enough parts may match our cognitive abilities. But the old vacuum tubes of first generation computers would have required a behemoth several times larger than New York even to match an australopithecine. Miniaturization is the key to revolution. With ever smaller and more compact silicon chips, computers, Jastrow claims, will soon reach human capacity at human sizes. What then, he asks in a final reverie, would prevent a mortal human from emptying the accumulated evolutionary and social experience of his mind into a machine and achieving electronic immortality? Might silicon-based intelligence, albeit with an organic, carbon-based helping hand at the outset, represent "the mature form of intelligent life in the Universe?"

I do not accept all of Jastrow's pronouncements on the merging of mind and machine, but I found this part of his book provocative and conceivably correct in outline. Unfortunately, it represents a small section at the end of a short book—an essay shorter than thirty pages, and not unsaid by others. The rest of the book attempts to portray this revolutionary development as the almost inevitable outcome of a continuous sequence of slow evolutionary improvements stretching back to the origin of life. As an essay in paleontology and evolutionary history, this discussion fails badly, and ends by portraying Jastrow's hopes and theology, rather than constraints of the world as biologists understand them.

Jastrow has tried to grant generality to his theme of computers transcending human intelligence by depicting the entire history of life as an inexorable and progressive march to increasing braininess; the transition from carbon to silicon then simply completes a universal directionality. We have not witnessed such a reincarnation of the old chain of being since Teilhard's Point Omega, and perhaps since Pope's *Essay on Man:*

> Mark how it mounts to man's imperial race
> From the green myriads in the peopled grass.

Ever since paleontology established the basic outlines of the fossil record more than a century ago, we have known how poorly the old chain of being matches the history of life. Its persistence as a metaphor and even, in Jastrow's case, as an imposed "reality" merely reflects our unwillingness to abandon comfort in the face of evidence.

I criticize two aspects of Jastrow's basic argument. First, even if life evolved as he states, this supposed directionality offers no guarantee of predictable continuity and advance in the transition from man to machine. Yet Jastrow draws this message from the history of life: "It is reasonable to assume that human beings are not the last word in the evolution of intelligence on earth. . . . The history of life supports this conclusion, for it shows a seemingly inexorable trend toward greater intelligence in the higher animals. . . . If the past is any guide to the future, mankind is destined to have a still more intelligent successor." But metaphor and analogy are not logical implication. Biological evolution is a theory about ties of physical genealogy based on reproduction with error and natural selection. Computers do not breed. Any direction imparted to biology by its Darwinian mechanism does not translate to pathways of industrial change; a biological past is no sure guide to a technological future.

Second, and more important, where is the "inexorable trend toward greater intelligence" that dominates Jastrow's biological vision? Most multicellular creatures are insects, doing very well thank you, and destined to outlive us, but not illustrating any temporal increase in intelligence to match their longstanding success. And each of our intestinal tracts contains more *E. coli* than the earth houses people. They will be with us at least until our intellectual essences enter those silicon chips. Life is a ramifying bush with millions of branches, not a ladder. Darwinism is a theory of local adaptation to changing environments, not a tale of inevitable progress. "After long reflection," Darwin wrote, "I cannot avoid the conviction that no innate tendency to progressive development exists."

Jastrow might argue that he is only considering the single pathway through the immense labyrinth of life's bush that happened to lead to us. Even here I might reply that while we have a personal motive for special interest in (and affection for) this particular pathway, we have no right to regard it (or any other) as *the* essential direction of life. The pathways leading to aardvarks, anchovies, or artichokes are just as long, intricate, and biologically informative.

Even if we grant Jastrow's special attraction for our pathway, where is the slow, smooth, and inexorable progress in intelligence that he sees? I don't wish to sound perverse. I don't deny that we started as single cells in the ocean, were once fish of limited intelligence, and can now build silicon chips, the agents of our future irrelevance. Still, we did not get from there to here on a ladder of ever increasing mental power. Each lineage is a series of curious accidents, with long periods of stability (or numerous variants on basic designs), and occasional unpredictable changes that, in retrospect, we sometimes choose to call progress. What if the ancestors of primates had died in the Cretaceous extinction, or dinosaurs survived? Would horses now be praising good legs, or tyrannosaurs sharp teeth, as a universal criterion of advance leading inexorably to domination? What if glaciers had frozen the whole earth, and not just a part?

Jastrow seems to know that the fossil records of our evolutionary lineage does not support his story of continuous and inexorable advance, so he tries to dig deeper by dubious inference. He struggles with the observation that nearly two-thirds of life's history belongs exclusively to bacteria and blue-green algae that look pretty much the same at the end of these three billion years as at the beginning. Jastrow, undaunted, talks about "some three billion years of invisible progress." In other cases, he distorts a story by ignoring diversity and only considering an abstracted archetype as the essence of a group. On supposed progress among the higher primates, he writes:

> The monkey did not change very much from the time of his appearance, 30 million years ago, to the present day. His story was complete. But the evolution of the ape continued. He grew large and heavy, and descended from the trees. . . .

What is *the* monkey? Monkeys come in more than one hundred species, from tiny marmosets in South America to baboons roaming the ground, to howler monkeys screaming and swinging through the trees. They are the most widespread and successful of primates. Apes are a dying afterthought (five species or so), with one peculiarly capable descendant. Moreover, with appropriate scaling to body weight, the small talapoin monkey has a larger brain than any ape.

Why is Jastrow's reading of life's history so different from that advanced by most people who study the fossil record for a living? Do I detect a theological bottom line? Jastrow and a few other astronomers have tried to find God in the universe by reading the big bang as the cosmological equivalent of Genesis. I confess that I have found it hard to take this argument seriously. The big bang may have created *our* universe. In addition, by disaggregating the products of any possible antecedent universe into basic particles, it may have obliterated the history of previous worlds. (This paradox is no different from Hutton's eighteenth-century contention that the cyclical nature of geological history wipes out the records of previous continents and oceans.) But an inability to reconstruct previous universes does not argue for their necessary nonexistence. We can only say that we do not know; the issue of whether the universe contains enough matter to contract again (pulsating versus unique big-bang theories) remains unresolved. If scientists should not play God, they should stop trying to find God as well. The inquiry may be legitimate, but not as a part of science.

Jastrow, nonetheless, persists in his quest,[1] and now wonders whether the supposedly inexorable progress of mentality does not point to a directing Intelligence. Is Paley's Watchmaker about to raise his head after a century of well-deserved slumber?

> When you study the history of life, and step back to look at this long history with the perspective of several hundred million years, you see a flow and a direction in it—from the simple to the complex, from lower forms to

[1]He also invokes the old red herring of a spurious conflict between the manifest order of life's history (which I do not deny, though I do not see it manifested as unilinear progress) and Darwinian theory characterized as an invocation of pure randomness. If Darwinism rests on pure chance, the argument goes, then it cannot explain the order of evolution and we require a (perhaps spiritual) ordering force. Jastrow writes: "If Darwin was correct, man has arisen on the earth as the product of a succession of chance events occurring during the last 4 billion years. Can that be true? Is it possible that man, with his remarkable powers of intellect and spirit, has been formed from the dust of the earth by chance alone?" The argument, though venerable, is entirely spurious and rests upon a basic misunderstanding of Darwin. The theory of natural selection invokes chance as a source of raw material only—genetic mutations. The preservation, accumulation and concatenation of favorable mutations—evolutionary *change* itself—are products of the deterministic and ordering force of natural selection. If, in the hoary metaphor, monkeys type at random, we will never get the *Aeneid* if we must start each trial from scratch. But if we may keep the letters that, by chance, turn up in the right places and start each new trial with these correct letters in place, we will eventually get "*Arma virumque cano*" and all the rest.

> higher, and always towards greater intelligence—and you wonder: Can this history of events leading to man, with its clear direction, yet be undirected?

But this "clear direction" is only an organizing principle in Jastrow's mind. To his vision, paleontologists can only reply with Christ's words to Pilate: "Thou sayest."

If the history of life is not a tale of smooth progress, then our putative transition from carbon to silicon intelligence cannot be linked with paleontology to form a unified and grandiose vision of continuity and advance in the universe. I would view such a switch from organism to machine as a true discontinuity in the earth's history. (I also willingly confess my personal predilection for viewing sharp transitions and changes of state as an important part of nature's panoply. What else does the output called "novelty" mean, however smooth the input?).

In one of his most famous aphorisms Freud argued for three great discontinuities in the history of Western science—each pushing humanity off a pinnacle of cosmic arrogance: first, the revolution of Copernicus, placing us not in the center of the universe but on "a tiny speck in a world-system of a magnitude hardly conceivable"; second, Darwin's revolution, which "robbed man of his peculiar privilege of having been specially created, and relegated him to a descent from the animal world"; third, in perhaps the least modest statement of intellectual history, Freud's own revolution, "endeavoring to prove to the 'ego' of each one of us that he is not even master of his own house, but that he must remain content with the veriest scraps of information about what is going on unconsciously in his own mind."

MIT historian Bruce Mazlish suggested in 1967 that computers would generate a "fourth discontinuity."[2] Each previous event had left us one comfort. Copernicus showed that we live on a peripheral hunk of rock, but we could still believe that God put us there by fiat. Darwin proved that we had evolved

[2] *Technology and Culture,* Vol. 8, No. 1.

naturally, but we still had our rational minds. Freud denied our rationality, but we could still view our mental power as unique. As the fourth pinnacle crumbles, we must admit that a board of silicon chips might surpass all the cognitive power in our heads.

But Mazlish also points out that each discontinuity in our own history establishes, as its cardinal substantive claim, an unperceived continuity in nature. First between the earth and other physical bodies, second between ourselves and nature, third between mind and evolved matter, and now, finally, between human life and the machines that we build. Ironically, Jastrow wrote this book to establish what I regard as a false continuity in the actual, physical evolution of intelligence. Yet the idea that artificial intelligence might unify nature may well be sound in a more abstract sense. By forging a true discontinuity in the physical history of intelligence on earth, we may force ourselves to appreciate our own deep embeddedness in nature. Of course, any paleontologist knows that too deep an embedding can lead to oblivion. This, indeed, is the paradox we may soon face.

16

Utopia, Limited

In his great work *An American Dilemma* (1944), Gunnar Myrdal praised the effort made by "a handful of social and biological scientists" to combat racism and hereditarianism—cultural prejudices once so pervasive that white intellectuals throughout the world had portrayed the inferiority of blacks as a self-evident truth. Myrdal then wondered what more general biases might lie so deep and unquestioned that we cannot even recognize them:

> But there must be still other countless errors of the same sort that no living man can yet detect, because of the fog within which our type of Western culture envelops us. Cultural influences have set up the assumptions about the mind, the body, and the universe with which we begin; pose the questions we ask; influence the facts we seek; determine the interpretation we give these facts; and direct our reaction to these interpretations and conclusions.

In *The Turning Point,* Fritjof Capra, author of *The Tao of Physics,* tries to identify and combat what may be the deepest bias of Western conceptual life, and a primary source (in his opinion) of our current ills and unhappiness. According to

A review of *The Turning Point: Science, Society, and the Rising Culture* by Fritjof Capra.

him, the trouble started nearly four hundred years ago when a previous organicism yielded to the conceptual paradigm of Cartesian mechanism—an approach pervading all disciplines and characterized by an explanatory tactic that separates, analyzes, and reduces the world to basic particles of atoms and molecules. This reductionist strategy has been allied with a conceptual machismo: belief in continual progress and growth by exploiting the earth and all its life (viewed therefore as separate from man and available for dominion), and the basic idea that we learn in order to control and manipulate ("knowledge is power," as Bacon proclaimed).

To halt our slide down this Cartesian path into an abyss of our own construction, Capra offers a new "paradigm of thought" now arising spontaneously, and often in unconscious or inchoate fashion, among troubled and perceptive thinkers in all disciplines. The watchwords of this new way are "holism" and "ecology." We must recognize inseparable union and interaction as basic realities. Complex systems, not separated building blocks, must be our units of explanation. We must immerse ourselves in nature and work in her ways, not separate ourselves from nature in order to exploit her.

Capra's book contains four parts. The first outlines the current crisis that Cartesianism has imposed upon us and offers some hints for an alternative. The second, in two chapters, sets forth the Cartesian model and then argues, by contrast, that its rationale has disappeared because advances in modern physics so well reflect the ecological and interactive themes of many non-Western and mystical traditions. (Capra, a physicist by training, here pursues several themes of his earlier book.) The third chronicles the dire influence of Cartesianism in biology, medicine, psychology, economics, and the politics of growth. The fourth proposes a holistic rescue in the same (but now indissoluble) areas.

This enormously right-minded general theme, here somewhat caricatured for brevity, surely wins my approval. My own recent work in evolutionary theory follows Capra's prescription: I have been trying to describe a hierarchical alternative to the Darwinian tradition that reduces all large-scale evolutionary phenomena to extrapolated results of natural selection working at the level of individual organisms within populations (the "struggle for existence," as Darwin stated, or, in modern terms, "differential reproductive success"). Hierarchical models by contrast, recognize genes, organisms, and species as legitimate entities in a sequence of levels with unique explanatory principles emerging at each more inclusive plateau.[1]

When King Paramount, in Gilbert and Sullivan's *Utopia Limited,* decides to transform his island paradise along modern British lines, he proclaims: "Though lofty aims catastrophe entail, we'll gloriously succeed, or nobly fail." *The Turning Point* is no catastrophe, but I do regard it as a noble failure, for two reasons. The first, not at all Capra's fault, harks back to Myrdal's insight: we are so embedded in Cartesian biases that we hardly know any other way to think. It is always easier to identify problems than to construct solutions. If Capra's description of the holistic and ecological paradigm lacks rigor and richness, well, many people are struggling with it—and no one has yet succeeded, so why should we expect more of him. The second, however, I do lay at his doorstep, for I find Capra's reasoning simplistic and even antirational (I think intentionally) at too many points. I shall concentrate, in turn, on the problems I see in his historical analysis of Cartesianism, his identification of the new paradigm, his intellectual justification for it, and his views on how we might enforce the substitution.

HISTORICAL ANALYSIS OF CARTESIANISM

The world is a complex place. In our struggles to simplify and understand, we often identify some bugbear and then make it responsible for all evils. Cartesian reductionism is Capra's

[1]S.J. Gould, "Darwinism and the expansion of evolutionary theory," *Science,* vol. 216 (1982), pp. 380–87; S.J. Gould, "The meaning of punctuated equilibrium and its role in validating a hierarchical approach to macroevolution," in: R. Milkman, ed., *Perspectives on Evolution* (Sinauer Assoc., Sunderland, MA, 1982), pp. 83–104; S.J. Gould & E.S. Vrba, "The Hierarchical expansion of sorting and selection: sorting and selection cannot be equated," *Paleobiology,* vol. 12 (1986), pp. 217–28.

candidate, and his far-fetched invocations of its baleful sway often seem ludicrous when much simpler explanations are available. Consider, for example, his account of why "in most European languages the right side is associated with the good, the just, the virtuous, the left side with evil, danger, and suspicion." Since the actions of our right side are mediated by the left hemisphere of our brain, and since the left hemisphere (in an oversimplified dichotomy so favored in scores of pop-psychology articles) performs "quantification and analysis" while the right thinks in holistic, integrated patterns, Capra argues that our preferences for right-handedness reflect "our culture's Cartesian bias in favor of rational thought." Has Capra forgotten Biblical preferences for him who "sitteth at the right hand of the father"—a bias originating in an age of pre-Cartesian organicism? Can he really pass by the obvious and simpler explanation for this pattern—that for some reason not yet understood most human beings are right-handed, and that good old xenophobia and fear of the unusual are quite sufficient to create our linguistic distinction of dexterous and sinister, without searching for Cartesian bugbears under every rug.

Moreover, Capra is so eager to blame the rise of Cartesianism for the origin of most Western problems that he paints an absurdly romantic view of a happy, holistic Europe before *Cogito, ergo sum.* For example,

> The value system that developed during the seventeeth and eighteenth centuries gradually replaced a coherent set of medieval values and attitudes—belief in the sacredness of the natural world; moral strictures against money-lending for interest; the requirement that prices should be "just"; convictions that personal gain and hoarding should be discouraged, that work was for the use value of the group and the well-being of the soul, that trade was justified only to restore the group's sufficiency, and that all true rewards were in the next world.

Tell it to the merchants of the Hanseatic League! I am not aware that a Europe so frequently ravaged by plague and war, with a peasantry so often oppressed and indentured, can lay any claim to such economic enlightenment.

My favorite scene in Brecht's *Galileo* dramatically paints the oppressive side of whatever holistic and organic elements resided in the paradigms of pre-Cartesian politics. The Little Monk, an astronomer who knows that Galileo is right, explains to his mentor why he will abandon the truth of the heavens and return to the Church's doctrine of a central earth. His parents, he says, are poor peasants in the Campagna and their lifelong suffering only makes sense if each creature plays a foreordained and inevitable role in the static harmony of permanent oneness. As the planets circle a central, controlling earth in a limited cosmos, so too must bishops defer to the Pope and peasants to their lords. Thus, says the Little Monk to Galileo, "Can you understand now that in the decree of the Holy Congregation I discern a noble motherly compassion, a great goodness of soul?" Galileo's reply is searing: "Damn it, I see the divine patience of your people, but where is their divine wrath?" Pre-Cartesian holism was more than a bucolic perception of nature's fundamental unity; it was also a dandy doctrine to enforce a status quo not blissful for everyone.

IDENTIFICATION OF THE NEW PARADIGM

As I have said, we are so embedded in the Cartesian world view that we hardly know how to formulate a general and coherent alternative. Capra himself quotes the perceptive biologist Sidney Brenner:

> I think in the next twenty-five years we are going to have to teach biologists another language. . . . I don't know what it's called yet; nobody knows. But what one is aiming at, I think, is the fundamental problem of the theory of elaborate systems. . . .

In the absence of a well-formulated substitute for Cartesian thinking, Capra is reduced to selective quotation from the heroes and harbingers of his new order. But buzz words and vague advocacy quickly pale into boredom. At best, we get hints from people who have worked out a holistic system only half way (von Bertalanffy), or in an oracular fashion (Gregory Bateson), or in the pop mode (Arthur Koestler). At worst, we have partial quotes from gurus who, so far as I can see, were not groping toward anything particularly anti-Cartesian, but whom Capra obviously wants in his pantheon. I fail to see, for example, why Teilhard de Chardin's harmless mystical reveries about the upward march of consciousness toward Point Omega should be seen as presaging the new holistic order simply because Teilhard once defined consciousness as " 'the specific effect of organized complexity,' which is perfectly compatible with the systems view of mind."

JUSTIFICATION OF THE NEW PARADIGM

When something hasn't been formulated rigorously, justification becomes a formidable task. As a basic strategy, Capra searches for similarity in the apparently disparate systems of people on the right track. If Western physicists and Eastern mystics are really saying the same thing in their ostensibly different struggles to understand deep reality, then, by God, there must be something to it. This is a dangerous and superficial approach to the analysis of similarity.

Natural historians, those scientists most directly charged with the task of analyzing similarity, have developed a system of classification both to sort kinds of meaning, and to divide the meaningful from the meaningless. Similarities are homologous, or genealogical, if they refer to structures retained by two or more objects from a common ancestor—my arm and a horse's forelimb, for example, with their markedly different functions, but striking similarity in bony structure. Homologous similarities gain a kind of intrinsic meaning as markers of a common inheritance.

Similarities developed independently and identified by formal or morphological resemblance are called analogous—wings of birds and bats, for example, since their common ancestor did not fly. Analogous similarities are particularly difficult to evaluate because they may be meaningful or meaningless depending upon context. Bird and bat wings tell us nothing about inheritance, but we may learn much about the aerodynamics of flight if a set of independent evolutionary events converge upon the same mechanical solution. When analogous similarities are regulated by the same physical constraints and laws, we can identify common reasons behind a unity of form even though each event occurred separately. The events may be quite diverse—hexagonal shapes of soap bubbles, plates on a turtle's shell, and basalt pillars in the Giant's Causeway of Northern Ireland, for example—but the same laws or formal causes may still underly the similarity, the geometric rules of closest packing in this case.

For analogies, however, an obvious problem arises because similarities not granted intrinsic meaning by genealogy may be striking in our subjective judgment but still quite accidental, random, superficial, or falsely perceived merely because we so desperately seek order in a confusing world. After all, the world contains far more objects than we have concepts—so we make such mistakes all the time. Surely most similarities fall into this superficial class, and we pass them by because our intuitions are well honed. (Some of us do not pass them by and tumble into all sorts of foolishness by, for example, listing common properties of Lincoln's and Kennedy's assassinations and believing that some deep reality has been revealed.)

It is always instructive to confront an older system of thought that granted deep meaning to similarities we now dismiss. Medieval bestiaries, for example, drew fanciful parallels between animals and their names, thinking that names themselves had independent meaning and must reveal not only the animals, but their deeper significance for our lives. My copy says that a goat is called *capra* in Latin (our author's name as well) because it strives to attain the mountain crags *(aspera* cap*tet)* and therefore represents Christ, who, in an anachronistic reference in the *Song of Songs,* "cometh like a he-goat leaping on the mountains."

Capra makes no attempt to distinguish meaningful from superficial similarities; he seems to feel that some organizing unity must lie behind whatever analogies happen to intrigue him. Thus his holistic paradigm emerges because so many different traditions that win his sympathy seem to be striving for the same oneness and simplification. But are they the same in any meaningful way?

As in his previous book, *The Tao of Physics,* Capra particularly stresses similarities between Eastern religious traditions and modern physics. For example, after laying out the dichotomy (and interpenetrating unity) of yin and yang, he decides that the complementarity of wave and particle descriptions for atomic phenomena records the same insight about reality constructed as an indissoluble system, not built from unambiguous items eventually "unpacked" at minute sizes.

But why should I accept this analogy as expressing a real unity in nature? Am I being too crudely analytical in noting that Chinese philosophers were not discussing basic particles even if they tried to construct a comprehensive system? Might not their yin and yang be reflecting our mind's struggle to grasp a complex reality by dichotomizing, rather than reordering nature herself? And why should I view the twofold nature of yin and yang as meaningfully similar to wave and particle in the first place? There are just so many ways to describe the world, and attempts often overlap without producing a eureka of true synthesis.

Further problems surround Capra's primary use of modern physics to support his holistic reconstruction. As the source of his illustrations, physics becomes his ultimate justification. He writes, for example:

> This is how modern physics reveals the basic oneness of the universe. It shows that we cannot decompose the world into independently existing smallest units. As we penetrate into matter, nature does not show us any isolated building blocks, but rather appears as a complicated web of relations between the various parts of a unified whole. . . .
>
> Here, at the level of particles, the notion of separate parts breaks down. The subatomic particles—and therefore, ultimately, all parts of the universe—cannot be understood as isolated entities but must be defined through their interrelations.

Consider the peculiarity of that last sentence: "the subatomic particles—and therefore, ultimately, all parts of the universe. . . ." The self-styled holist and antireductionist is finally caught in his own parochialism after all. He has followed the oldest of reductionist strategies. As it is with the structure of physics, queen of the sciences, so must it be, by extrapolation, with all of nature. You don't exit from this Cartesian trap by advocating holism at the lowest level. The very assertion that this lowest level, whatever its nature, represents the essence of reality, *is* the ultimate reductionist argument.

The hierarchical perspective must take seriously the principle that phenomena of one level cannot automatically be extrapolated to work in the same way at others (a stricture that must be treated with special respect when we indulge our Cartesian habits for explaining the large in terms of the small). It might well be (though I doubt it) that subatomic phenomena are indissoluble, while houses, sequoias, and rhinos are wondrously distinct. And both will represent equally valid aspects of reality.

Finally, I find that too much of Capra's supposed justification for holism rests upon a simple glorification of the nonrational. We encounter, for example, the persistent theme that nature's way must be right and that our bumbling minds err if we do anything different. This argument is particularly ill-suited when Capra's notion of nature is so wanting:

> By developing our capacity for abstract thinking at such a rapid pace, we seem to have lost the important ability to ritualize social conflicts. Throughout the animal world aggression rarely develops to the point where one of the two adversaries is killed. Instead, the fight is ritualized and usually ends with the loser conceding defeat but remaining relatively unharmed. This wisdom disappeared, or at least was deeply submerged, in the emergent human species. In the process of creating an abstract inner world we seem to have lost touch with the

> realities of life and have become the only creatures who often fail to cooperate with and even kill their own kind.

This idea of ritualized combat may have been a transiently popular theme in ethology fifteen years ago, but recent studies of animal behavior have uncovered, in species after species, far more injury and death and far less abstract posturing than we once imagined. So be it. Nature has no automatically transferable wisdom to serve as the basis of human morality. Passive observation and unquestioned reverence for nature are no substitute for ethical philosophy.

In other passages, the blessings of holism (even its justification) are equated with those emotional rushes that cut through reason and make us so sure that all is one:

> At rare moments in our lives we may feel that we are in synchrony with the whole universe. These moments may occur under many circumstances—hitting a perfect shot at tennis or finding the perfect run down a ski slope, in the midst of a fulfilling sexual experience, in contemplation of a great work of art, or in deep meditation.

I don't abjure these joys, especially the third and fourth. But why must *all* Capra's examples lie so pointedly in the realm of the nonrational? Am I so peculiar because some of my greatest emotional highs have accompanied my understanding of a bit of nature's complexity? And why, oh why, must the athletic examples always be tennis and skiing? Because this is what the beautiful people do? Is anything really wrong with softball and bowling, my own persistent favorites, despite their odor of ordinariness?

I hate to say this, because it reveals an unwarranted parochialism on my own part. I thought that Capra and I would be kindred spirits, since we maintain a similar commitment to a holistic and hierarchical perspective. Yet I found myself getting more and more annoyed with his book, with its facile analogies, its distrust of reason, its invocation of fashionable notions. In some respects, I feel closer to rational Cartesians (at least we have a common basis for disagreement) than to Capra's California brand of ecology. I guess I'm just a New York holist.

HOW TO MAKE THE TRANSITION

If Cartesianism is so pervasive a basis for scientific work, and if it grants such benefits to those in power, then it will not be happily abandoned in an orgy of expiation. Those who reap the spoils will not easily surrender them. Capra, however, seems little troubled by political realities and does not probe much beyond the faith that transitions so desirable are somehow bound to happen. "Following the philosophy of the I Ching rather than the Marxist view," he writes, "I believe that conflict should be minimized in times of social transition." Nice if you can have it. Perhaps, he suggests, holistic pressure groups will make the desirability of transition so apparent that former nonvoters will flock to the polls and "turn the paradigm shift into political reality."

I find it both depressing and amusing that so many of our intellectual efforts, though masquerading as attempts to understand nature, are really anodynes for justifying our hopes and calming our fears. We have had central earths in small universes and people created in God's physical image. Capra's version of the holistic vision ultimately fails because it follows the same tradition and substitutes desire for analysis. From the political theme that good things will happen because they should, to the search for analogies that must be meaningful because admirable people present them from different worthy traditions, Capra's claims and prognoses often do little more than promote his personal preferences as nature's way.

Most seriously, I think that Capra's focus on his own hopes has led to a basic misunderstanding of the very paradigm that he advocates. The antireductionist model tries to understand nature as a hierarchy of interdependent levels, each coherent in itself, but each linked by ties of feedback to adjacent levels (Arthur Koestler's Janus-faced holon, for example, who stands firmly in one spot but looks both up and down at the same time). No level is an ultimate reality and reference point for extrapolation; all are legitimate, interacting aspects of our natural world.

Capra, in his hopes for harmony and oneness, continually argues that the levels must reinforce one another and lead, ultimately, to the same desired end for all. He writes, for example:

> Accordingly, the systems view of health can be applied to different systems levels, with the corresponding levels of health mutually interconnected. In particular we can discern three interdependent levels of health—individual, social, and ecological. What is unhealthy for the individual is generally also unhealthy for the society and for the embedding ecosystem.

Would it were so. The world would be an easier place if this hope of interacting harmony followed inevitably from a concept of levels. But it doesn't. If levels have substantial independence, then advantages to individuals at one level may or may not yield benefits to individuals at adjacent levels. Blessings to persons need not benefit collectivities—unless we are trying to graft our hopes for harmony upon an ultimately uncooperative world.

The benefits of adjacent levels are often contradictory; "harmonious" solutions must be balances that satisfy neither level completely, not the inevitable discovery of a pathway optimal for all. In my own field of evolutionary biology, benefits to organisms in the immediate struggle for existence often (probably usually) reduce the longevity of species (next level up). Intricate and complex structures, from peacocks' tails to the elaborate armature of fighting males, confer evolutionary success upon the biggest and brightest organisms, but commit a species to such extreme specialization for particular environments that it will perish when conditions change—while related species that remain small and drab may retain enough evolutionary flexibility to weather the change. Sports heroes are not hurting themselves with seven-figure salaries, but the economic structure of the games they play may be threatened thereby.

Holism does not imply necessary harmony; conflicts among legitimate demands of different levels are as much a part of the hierarchical model as confluence. We may develop a new paradigm, as Capra believes, but it will not produce a nirvana of self-benefiting cooperation. The world remains too complex. I see no intrinsic bar to a decent life for all, but I doubt that we will ever escape sacrifice, struggle, and compromise.

17

Integrity and Mr. Rifkin

Evolution has a definite geometry well portrayed by our ancient metaphor, the tree of life. Lineages split and diverge like the branches of a tree. A species, once distinct, is permanently on its own; the branches of life do not coalesce. Extinction is truly forever, persistence a personal odyssey. But art does not always imitate nature. Biotechnology, or genetic engineering, has aroused fear and opposition because it threatens to annul this fundamental property of life—to place genes of one species into the program of another, thereby combining what nature has kept separate from time immemorial. Two concerns—one immediate and practical, the other distant and deep—have motivated the opposition.

Some critics fear that certain conjunctions might have potent and unanticipated effects—creating a resistant agent of disease or simply a new creature so hardy and fecund that, like Kurt Vonnegut's *ice-nine,* it spreads to engulf the earth in a geological millisecond. I am not persuaded by these excursions into science fiction, but the distant and deeper issue does merit discussion: What are the consequences, ethical, aesthetic, and practical, of altering life's fundamental geometry and permitting one species to design new creatures at will,

A review of *Algeny* by Jeremy Rifkin, from *Discover,* January 1985. All other essays in this volume originally appeared in the *New York Review of Books.*

combining bits and pieces of lineages distinct for billions of years?

Jeremy Rifkin has been our most vocal opponent of genetic engineering. He has won court cases and aroused fury in the halls of science with his testimony about immediate dangers. However, his major statement, a book titled *Algeny* (for the modern alchemy of genes), concentrates almost entirely on the deep and distant issue. His activities based on immediate fears have been widely reported and rebutted. But *Algeny* has not been adequately analyzed or dissected. Its status as prophecy or pretension, philosophy or pamphleteering, must be assessed, for *Algeny* touts itself as the manifesto of a movement to save nature and simple decency from the hands of impatient and rapacious science.

I will state my conclusion—bald and harsh—at the outset. I regard *Algeny* as a cleverly constructed tract of anti-intellectual propaganda masquerading as scholarship. Among books promoted as serious intellectual statements by important thinkers, I don't think I have ever read a shoddier work. Damned shame, too, because the deep issue is troubling and I do not disagree with Rifkin's basic plea for respecting the integrity of evolutionary lineages. But devious means compromise good ends, and we shall have to save Rifkin's humane conclusion from his own lamentable tactics.

The basic argument of *Algeny* rests upon a parody of an important theme advanced by contemporary historians of science against the myth of objectivity and inexorable scientific progress: science is socially embedded; its theories are not simple deductions from observed facts of nature, but a complex mixture of social ideology (often unconsciously expressed) and empirical constraint. This theme is liberating for science; it embodies the human side of our enterprise and depicts us as passionate creatures struggling with limited tools to understand a complex reality, not as robots programmed to convert objective information into immutable truth. But in Rifkin's hands, this theme becomes a caricature. Rifkin ignores the complex interplay of social bias with *facts* of nature and promotes a crude socioeconomic determinism that views our historical succession of biological world-views—from creationism to Darwinism to the new paradigm now supposedly under construction—as so many simple reflections of social ideology.

From the crudity of this socioeconomic determinism, Rifkin constructs his specific brief: Darwinian evolutionism, he asserts, was the creation of industrial capitalism, the age of pyrotechnology. Arising in this context as a reflection of social ideology, it never had any sound basis in reason or evidence. Darwinism is now dying because the age of pyrotechnology is yielding to an era of biotechnology—and biotech demands a new view of life. Darwinism translated the industrial machine into nature; biotech models nature as a computer and substitutes information for material parts.

Darwinism spawned (or reflected) evil in its support for exploitation of man and nature, but at least Darwinism respected the integrity of species (while driving some to extinction) because it lacked the technology to change them by mixture and instant transmutation. But the new paradigm dissolves species into strings of information that can be reshuffled at will.

The new temporal theory of evolution replaces the idea of life as mere machinery with the notion of life as mere information. All living things are drained of their aliveness and turned into abstract messages. We can no longer entertain any question of sacredness or inviolability. What could such concepts mean in a world with no recognizable boundaries to respect? In the age of biotechnology, separate species with separate names gradually give way to systems of information that can be reprogrammed into an infinite number of biological combinations.

But what can we do if we wish to save nature as it actually evolved—a system divided into packages of porcupines and primroses, cabbages and kings? We can seek no help from science, Rifkin claims, for science is a monolith masquerading as objective knowledge, but really reflecting the dominant ideology of a new technological age. We can only make an ethical decision to "re-sacralize" nature by respecting the inviolability of its species. We must, for the first time in history, decide *not* to institute a possible technology, despite its immediately attractive benefits in such areas as medicine and agriculture.

I have devoted my own career to evolutionary biology, and I have been among the strongest critics of strict Darwinism. Yet Rifkin's assertions bear no relationship to what I have observed and practiced for 25 years. Evolutionary theory has never been healthier or more exciting. We are experiencing a ferment of new ideas and theories, but they are revising and extending Darwin, not burying him. How can Rifkin construct a world so different from the one I inhabit and know so well? Either I am blind or he is wrong—and I think I can show, by analyzing his slipshod scholarship and basic misunderstanding of science, that his world is an invention constructed to validate his own private hopes. I shall summarize my critique in five charges:

1. Rifkin does not understand Darwinism, and his arguments refute an absurd caricature, not the theory itself. He trots out all the standard mischaracterizations, usually confined nowadays to creationist tracts. Just three examples: "According to Darwin," Rifkin writes, "everything evolved by chance." Since the complexity of cellular life cannot arise by accident, Darwinism is absurd: "According to the odds, the one-cell organism is so complex that the likelihood of its coming together by sheer accident and chance is computed to be around $1/10^{78438}$." But Darwin himself, and Darwinians ever since, have always stressed, as a cardinal premise, that natural selection is not a theory of randomness. Chance may describe the origin of new variation by mutation, but natural selection, the agent of change, is a conventional deterministic process that builds adaptation by preserving favorable variants.

Rifkin then dismisses Darwinism as a tautology; fitness is defined by survival, and the catch phrase "survival of the fittest" reduces to "survival of those that survive"—and therefore has no meaning. Darwin resolved this issue by defining fitness as predictable advantage before the fact, not as recorded survival afterward (as we may predict the biomechanical improvements that might help zebras outrun or outmaneuver lions; survival then becomes a testable consequence of good design).

Rifkin regards Darwinism as absurd because "natural selection makes no room for long-range considerations. Every new trait has to be immediately useful or it is discarded." How, therefore, can natural selection explain the origin of a bird's wing, since the intermediate forms cannot fly: What good is five per cent of a wing? The British biologist St. George Jackson Mivart developed this critique in 1871 as the argument about "incipient stages of useful structures." Darwin met the challenge by adding a chapter to the sixth edition of the *Origin of Species.* One need not agree with Darwin's resolution, but one does have a responsibility to acknowledge the argument's existence. Darwin proposed that intermediate stages performed different functions; feathers of an incipient wing may act as excellent organs of thermoregulation—a particular problem in the smallest of dinosaurs, the lineage that evolved into birds.

Rifkin displays equally little comprehension of basic arguments about evolutionary geometry. He thinks that *Archaeopteryx* has been refuted as an intermediate link between reptiles and birds because some true birds have been found in rocks of the same age. But evolution is a branching bush, not a ladder. Ancestors survive after descendants branch off. Dogs evolved from wolves, but wolves (though threatened) are hanging tough. And a species of *Australopithecus* lived side by side with its descendant *Homo* for more than a million years in Africa.

Rifkin doesn't grasp the current critiques of strict Darwinism any better. He caricatures my own theory of punctuated equilibrium as a sudden response to ecological catastrophe: "The idea is that these catastrophic events spawned monstrous genetic mutations within existing species, most of which were lethal. A few of the mutations, however, managed to survive and become the precursors of a new species." But punctuated equilibrium, as Niles Eldredge and I have always emphasized, is a theory about ordinary speciation (taking tens of thousands of years) and its abrupt appearance at low scales of geological resolution, not about ecological catastrophe and sudden genetic change.

Rifkin, it appears, understands neither the fundamentals of Darwinism, its current critiques, nor even the basic topology of the evolutionary tree.

2. Rifkin shows no understanding of the norms and procedures of science: he displays little comprehension of what science is and how scientists work. He consistently misses the essential distinction between fact (claims about the world's empirical content) and theory (ideas that explain and interpret facts)—using arguments from one realm to refute the other. Against Darwinism (a theory of evolutionary mechanisms) he cites the British physiologist Gerald Kerkut's *Implications of Evolution,* a book written to refute the factual claim that all living creatures have a common ancestry, and to argue instead that life may have arisen several times from chemical precursors—an issue not addressed by Darwinism. (Creationist lawyers challenged me with the same misunderstanding during my cross-examination at the Arkansas "equal time" trial five years ago.) Rifkin then suggests that the entire field of evolution may be pseudo science because the great French zoologist Pierre-Paul Grassé is so critical of Darwinism (the theory of natural selection might be wrong, but Grassé devoted his entire life to studying the facts of evolution).

Science is a pluralistic enterprise, validly pursued in many modes. But Rifkin ignores its richness by stating that direct manipulation by repeatable experiment provides the only acceptable method for reaching a scientific conclusion. Since evolution treats historically unique events that occurred millions of years ago, it cannot pass muster. Rifkin doesn't seem to realize that he is throwing out half of science—nearly all of geology and most of astronomy, for instance—with his evolutionary bath water. Historical science is a valid pursuit, but uses methods different from the controlled experiment of Rifkin's all-encompassing caricature—search for an underlying pattern among unique events, and retrodiction (predicting the yet undiscovered results of past events), for example.

3. Rifkin does not respect the procedures of fair argument.

He uses every debater's trick in the book to mischaracterize and trivialize his opposition, and to place his own dubious claims in a rosy light. Just four examples:

The synecdoche (trying to dismiss a general notion by citing a single poor illustration). Rifkin suggests that science knows nothing about the evolutionary tree of horses, and has sold the public a bill of goods (the great horse caper, he calls it), because one exhibit, set up at the American Museum of Natural History in 1905, arranged fossil horses in order of size, not genealogy. Right, Jeremy, that was a lousy exhibit, but you might read George Gaylord Simpson's book *Horses* to see what we do know.

The half quote (stopping in the middle so that an opponent appears to agree with you, or seems merely ridiculous). Rifkin quotes me on the argument about incipient stages of useful structures discussed a few paragraphs ago: "Harvard's Stephen Jay Gould posed the dilemma when he observed, 'What good is half a jaw or half a wing?' " Sure, I posed the dilemma, but then followed it with an entire essay supporting Darwin's resolution based on different function in intermediate stages. Rifkin might have mentioned the true subject and not cited me in his support. Rifkin then quotes a famous line from Darwin as if it represented the great man's admission of impotence: "Darwin himself couldn't believe it, even though it was his own theory that advanced the proposition. He wrote: 'To suppose that the eye, with all of its inimitable contrivances . . . could have been formed by natural selection, seems, I freely confess, absurd in the highest possible degree.' " But Rifkin might have mentioned that Darwin follows this statement with one of his most brilliant sections—a documentation of nature's graded intermediates between simple pinhole eyes and the complexity of our own, and an argument that the power of new theories resides largely in their ability to resolve previous absurdities.

Refuting what your opponents never claimed. In the 1950s, Stanley Miller performed a famous experiment that synthesized amino acids from hypothetical components of the earth's original atmosphere. Rifkin describes Miller's experiment with glaring hype: "With great fanfare, the world was informed that scientists had finally succeeded in forming life from non-life, the dream of magicians, sorcerers, and alchemists from time immemorial." He then points out, quite correctly, that the experiment did no such thing, and that the distance from amino acid to life is immense. But Miller never claimed that he had made life. The experiment stands in all our textbooks as a demonstration that some simple components of living systems can be made from inorganic chemicals. I was taught this interpretation 25 years ago; I have lectured about this experiment for 15 years. I have never in all my professional life heard a scientist say that Miller or anyone else has made life from non-life.

Refuting what your opponents refuted long ago. Rifkin devotes a whole section to ridiculing evolution because its supporters once advanced the "biogenetic law" that embryos repeat the adult stages of their ancestry—now conclusively refuted. But Darwinian evolutionists did the refuting more than 50 years ago (good science is self-correcting).

4. Rifkin ignores the most elementary procedures of fair scholarship. His book, touted as a major conceptual statement about the nature of science and the history of biology, displays painful ignorance of its subject. His quotations are primarily from old and discredited secondary sources (including some creationist propaganda tracts). I see no evidence that he has ever read Darwin in the original. He obviously knows nothing about (or chooses not to mention) all the major works of Darwinian scholarship written by modern historians. His endless misquotes and half quotes are, for the most part, taken directly from excerpts in hostile secondary sources.

His prose is often purple in the worst journalistic tradition. When invented claims are buttressed by such breathless description, the effect can be highly amusing. He mentions T.H. Morgan's invocation of the tautology argument discussed previously in this essay: "Not until Morgan began to suspect that natural selection was a victim of circular reasoning did anyone in the scientific community even question what was regarded by all as a profound truth . . . Morgan's observation shocked the scientific establishment." Now, I ask, how does he know this? Rifkin cites no evidence of any shock, even of any contemporary comment. He quotes Morgan himself only from secondary sources. In fact, everything about Rifkin's statement is wrong, just plain wrong. The tautology argument dates from the 1870s. Morgan didn't invent it (and Darwin, in my opinion, presented an able refutation when Morgan was a baby). Morgan, moreover, was no noble knight sallying forth against a monolithic Darwinian establishment. When Morgan wrote his critique, natural selection was an unpopular minority theory among evolutionists (the tide didn't turn in Darwin's favor until the late 1930s). Morgan, if anything, *was* the establishment, and his critique, so far as I know, didn't shock a soul or elicit any extensive commentary.

5. *Algeny* is full of ludicrous, simple errors. I particularly enjoyed Rifkin's account of Darwin in the Galapagos. After describing the "great masses" of vultures, condors, vampire bats, and jaguars, that Darwin saw on these islands, Rifkin writes: "It was a savage, primeval scene, menacing in every detail. Everywhere there was bloodletting, and the ferocious, unremittent battle for survival. The air was dank and foul, and the thick stench of volcanic ash veiled the islands with a kind of ghoulish drape." Well, I guess Rifkin has never been there; and he obviously didn't bother to read anything about these fascinating islands. None of those animals live on the Galapagos. In fact, the Galapagos house no terrestrial predators at all; as a result, the animals have no fear of humans and do not flee when approached. The Galapagos are unusual, as Darwin noted, precisely because they are *not* scenes of Hobbes's *bellum omnium contra omnes* (the war of all against all). And, by the way, no thick stench or ghoulish drape either; the volcanic terrains are beautiful, calm, and peaceful—not in eruption when Darwin visited, not now either.

Jeremy Rifkin, in short, has argued himself, inextricably, into a corner. He has driven off his natural allies by silly, at times dishonest, argument and nasty caricature. He has saddled his legitimate concern with an extremism that would outlaw both humane and fascinating scientific research. His legitimate brief speaks for the integrity of organisms and species. Our world would become a bleak place if we treated living things as no more than separable sequences of information, available for disarticulation and recombination in any order that pleased human whim. But I do not see why we should reject all genetic engineering because its technology might, one day, permit such a perversion of decency in the hands of some latter-day Hitler—you may as well outlaw printing because the same machine that composes Shakespeare can also set *Mein Kampf.* The domino theory does not apply to all human achievements. If we could, by transplanting a bacterial gene, confer disease or cold resistance upon an important crop plant, should we not do so in a world where people suffer so terribly from malnutrition? Must such a benefit imply that, tomorrow, corn and wheat, sea horses and orchids will be thrown into a gigantic vat, torn apart into genetic units, and reassembled into rows of identical human servants? Eternal vigilance, to recombine some phrases, is the price of technological achievement.

The debate about genetic engineering has often been portrayed, falsely, as one of many battles between the political left and right—leftists in opposition, rightists plowing ahead. The issues are not so simple; they rarely are. Used humanely for the benefit of ordinary people, not the profits of a few entrepreneurs, the left need not fear this technology. I, for one, would rather campaign for proper use, not abolition. If Rifkin's argument embodies any antithesis, it is not left versus right, but romanticism, in its most dangerous anti-intellectual form, versus respect for knowledge and its humane employment. In both its content and presentation, *Algeny* belongs in the sordid company of anti-science. Few campaigns are more dangerous than emotional calls for proscription rather than thought.

I have been so harsh because I believe that Rifkin has seriously harmed a cause that is very dear to me and to nearly all my scientific colleagues. Rifkin has placed all of us beyond the pale of decency by arguing that scientific paradigms are simple expressions of socioeconomic bias, that biotech implies (and will impose) a new view of organisms as strings of separable information (not wholes of necessary integrity), and that all scientists will eventually go along with this heartless idea. Well, Mr. Rifkin, who then will be for you? Where will you find your allies in the good fight for respect of evolutionary lineages? You have rejected us, reviled us, but we are with you. We are taxonomists, ecologists, and evolutionists—most of us Darwinians. We have devoted our lives to the study of species in their natural habitats. We have struggled to understand—and we greatly admire—the remarkable construction and operation of organisms, the product of complex evolutionary histories, cascades of astounding improbability stretching back for millions of years. We know these organisms, and we love them—as they are. We would not dissolve this handiwork of four billion years to satisfy the hubris of one species. We respect the integrity of nature, Mr. Rifkin. But your arguments lack integrity. This we deplore.

18

The Quack Detector

In the heady days, only a decade ago, when psychologists thought they had unlocked the conceptual capacities of apes by teaching them American Sign Language, a leading researcher confessed to me that he would refrain from teaching one key item to his chimpanzee—the fact of her impending personal mortality. No other animal, he explained, understood this most terrible of all facts—and he had nightmarish visions of his enlightened ape spreading the bad news by sign throughout chimpdom.

Ever since we learned this fact as the most unfortunate consequence of evolving a larger brain, we have done our best to mitigate its doleful message. I remembered this recently when I sang Bach's great motet *Jesu meine Freude* and came to that sublime fugue with the most God-awful tongue-twisting text: "*Sie aber seid nicht fleischlich, sondern geistlich*"—you are not made of flesh, but of spirit. Much of the greatest work in philosophy, religion, art, and music exists either to bewail our mortality or to argue that a spiritual continuity permits us to accept the physical decline and eventual decay of our bodies.

Long before P.T. Barnum drew the correct equation between the birth of suckers and the passage of minutes, this legitimate search for mitigation had its counterpart in a vast nether world of huckstering and nonsense about occult phenomena. We either seek to communicate directly with

A review of *Science: Good, Bad and Bogus* by Martin Gardner.

souls in the beyond (spiritualism) or merely in the present (telepathy and other forms of ESP), or we invent an independent, higher realm of spiritual forces and hope that we can plug in by harnessing its powers (psychokinesis) or living according to its laws (astrology).

Moreover—and by what elitist arrogance should we think otherwise—occultism has always been as fashionable in chic intellectual circles as in drugstore paperbacks and *The National Enquirer.* Several years ago, I wrote to the manager of the Harvard Coop bookstore, complaining that their paperback science section had been moved to a less visible position on another floor and replaced with a large section on astrology and the occult. He replied that science had not been eliminated, merely moved to reflect a "sales reality." I replied that I had never doubted the reason, but had written to protest. We had clearly reached an impasse.

In this climate, beleaguered rationalism needs its skilled debaters—writers who can combine wit, penetrating analysis, sharp prose, and sweet reason into an expansive view that expunges nonsense without stifling innovation, and that presents the excitement and humanity of science in a positive way, not (to quote the immortal words of Mr. Agnew, via Mr. Safire) like a "nattering nabob of negativism."

For more than thirty years, Martin Gardner has played this largely thankless role with tireless efficiency and rarely strained good humor. He is more than a mere individual fighting a set of personal battles; he has become a priceless national resource. Since resources demand anthologies, I welcome this collection of Gardner's writings from 1950 to 1980.

Our will to believe forms the substrate that nurtures literature of the occult. This yearning for easy meaning or immortality cannot be overcome by logical debunking. But occultism also gains support from two more specific sources that can be effectively attacked. Gardner works in these arenas of possible success.

First, human gullibility has cash value, and enormous amounts of money can be made by any skilled manipulator. For every sincere (however naïve) researcher in parapsychology, ten charlatans misuse the legitimate art of stage magic to enhance prestige and pocketbook. Real magicians, from the great Houdini in decades past to the Amazing Randi today, have laboriously exposed the data of ESP as tricks of their profession. Yet the will to believe is unbounded, and true disciples merely reply that, although their mentor fakes sometimes, his real spiritual powers exist nonetheless. People will believe the damnedest things. Arthur Conan Doyle wrote an entire book on the existence of fairies, holding firm even when his best case had been exposed as photographs of crude cardboard cutouts. (Gardner, in an amusing essay, suggests that Sherlock Holmes would not have permitted such a man to write his memoirs.) Still, true believers aside, the exposure of massive and persistent fakery is a strong argument, at least for skeptical caution.

Second, general gullibility is often greatly enhanced among scientists by an arrogance leading some to proclaim that a person trained in observation and experiment should be able to decide whether any man wields true psychic power or performs clever stage magic. Most ordinary mortals respect the art of magicians and are prepared to be fooled. (I cannot decipher the simplest card trick and place myself firmly in this company.) Some scientists feel that their skills will detect any fakery—and they can really be fooled. Gardner, an accomplished amateur magician himself, shows how the Uri Gellers of this world use stock stage magic (not even with particularly great skill) to make some arrogant scientists a laughing stock of the magicians' fraternity and, unfortunately, a stalking-horse for irrationality in the guise of simple fakery.

Since we scientists are forever demanding deference to our professional skills, we could at least respect other equally exacting crafts, and not look down upon them because they thrive on the stage, but not in the academy. If every parapsychologist followed the simple rule of always including a professional magician in any test of people claiming extrasensory powers, millions of dollars, thousands of hours, and hundreds of reputations would be saved. Similarly, if the psychologists who tried to teach sign language to chimpanzees had bothered to consult the real professionals in this area—the great animal trainers of our major circuses—they might have avoided some spectacular (and now spectacularly embarrassing) claims for conceptualization and consciousness that now seem to arise from unconscious human cueing and simple coincidence.

Walt Whitman exhorted us to "make much of negatives, and of mere daylight and the skies." It is easy to debunk from the empyrean platform of established science, to be haughtily, exclusively, and uncompromisingly negative. The challenge lies in preserving daylight in the midst of excoriation, for it cannot be said often enough that quacks grade to cranks and cranks to geniuses through the finest intermediary stages. All enlightened debunkers must bear this cross: to be ever open to honest nuttiness while ruthlessly exposing the frauds, yet to be accused by all opponents (however falsely) of being the pawn for an oppressive establishment trying to hide tumultuous truths from a thirsting public.

The expansive nature of Gardner's debunking is best demonstrated in the book's finest essay—a treatise on the *Ars magna* (Great Art) of the thirteenth-century Catalan mystic Ramon Lull. Lull tried to dissolve the distinction between theological and philosophical truth and to demonstrate that even the deepest mysteries of Christianity could be proved by logical argument. He developed a system and a set of geometric devices for generating all possible combinations among sources of truth. His age, for example, recognized seven virtues, seven vices, and seven planets (sun, moon, and five visible planets of an older cosmology). Lull therefore constructed a wheel of three concentric circles, each divided into seven equal parts and each free to rotate about their common center. All possible combinations of the seven items taken three at a time can be easily generated by rotating the wheels into all positions. As Gardner states, Lull believed that "by exhausting the combinations of such principles one might explore all possible structures of truth and so obtain universal knowledge."

Lull probably transgressed the boundary between insight and crankiness; his later dogmatic disciples certainly did. It is easy to generate the combinations, but who can read their unambiguous meaning? Still, since tangential thinking by combination of unexpected items may be a more important component of creativity than logical deduction, Lull's methods have much to teach us, and his little machines may even have their uses. The importance of Lull, Gardner emphasizes, lies not in his own excursion around the bend, but in his honest and groping struggle toward unusual forms of legitimate insight.

Gardner's book is not without its problems. Since effective debunking demands constant and patient repetition, some items cycle through the essays a few too many times. To this criticism I must add the Catch-22 of Gardner's art: victory renders the specific subject irrelevant. Yesterday's seer is today's bore. Who cares about Uri Geller since we all now know (I trust) that he is a skilled con man and a mediocre magician. Geller once is a good reminder; Geller by the dozen begins to wear.

Moreover, although I don't blame Gardner for obeying the unwritten rules of his magicians' guild, it is frustrating to be told that somebody fooled a bunch of eminent scientists with a simple trick known to all professionals, and then to be put off by a gentle reminder that magicians never tell. I could understand if magic were an arcane art, rigidly regulated by its devotees, and never revealed to outsiders on pain of permanent ostracism or death. But "how to" books and pamphlets abound, though few of us have them on our shelves for ready reference while reading this book. Gardner might have relieved our persistent frustration by being a bit more forthcoming.

Finally, on a smaller item in the "oughta be a law" category, may I bemoan the lack of an index. In the bad old days, the index was a list of prohibited books; may we now, in a more enlightened age, ban books without indexes?

We all applaud the unmasking of scoundrels and take a voyeuristic delight in the exposed foolishness of our fellows. Still, many people probably regard the debunking of occult and paranormal claims as a marginal exercise—even while they must admit that most people encounter planets more often in horoscopes than in planetariums or (wonder of wonders) in the sky itself. I would cite two reasons for regarding the failure of critical faculties and the decision to accept improbable hope as rather more tragic than merely amusing.

First, lives are short and resources generally slim. It may be their own fault, but what do people say when they wake up twenty years and half a career later to acknowledge that they have wasted their time on a chimera nurtured by fraud? And what of people who invest their deepest hopes (and much of their hard cash) on harebrained schemes for spiritual enlightenment or continuity (the wisdom of psychic forces or little green folks in UFOs, or direct information from Uncle George in the beyond)?

Moreover, the density of fraud and nonsense in parapsychology drives critical and discerning scientists away from a subject that may display chinks of enormous promise. So-called ESP, after all, is not impossible a priori (as Gardner continually acknowledges), but who wants to invest precious years of a career in an area so rife with fakery not easily detected by the ordinary methods of science? (My snails hide their secrets, but they don't lie—and I wouldn't know how to unmask them if they did.) Thus, ironically, the fools and frauds are keeping their own ship from a potential port. As one colleague put it to me: suppose a man with eyes were born to a race of blind men. He comes to a new village and asks its inhabitants, "Is the water in that pond half a mile yonder fit to drink?" "How do you know the pond is there?" they inquire in disbelief. "I see it," the man replies. There may be a physics of rare skills, but how can we discern it amid the quackery?

Second, as we discern a fine line between crank and genius, so also (and unfortunately) must we acknowledge an equally graded trajectory from crank to demagogue. When people learn no tools of judgment and merely follow their hopes, the seeds of political manipulation are sown. Consider the current bugbear of my own profession—resurgent creationism. Some creationist beliefs are so downright ridiculous that we might be tempted, at our great peril, to dismiss them with laughter. Just consider, for example, the so-called "flood geology" espoused by nearly all professional creationists in America today—the claim that all geological strata, with their exceptionless, worldwide sequence of fossils, are products of a single event: Noah's universal flood and its resultant fallout. Why, then—no place anywhere on this vast earth—do we find dinosaurs and large mammals in the same strata; why are trilobites never with mammals, but always in strata below? One might argue that dumb dinosaurs were less skilled at avoiding flood waters than bright mammals, and got buried earlier. One might claim that trilobites, as denizens of the ocean, were entombed before terrestrial mammals. But why are they never found with whales? Surely some retarded elephant would be keeping company with dinosaurs, some valiant trilobite swimming hard for thirty-nine days and winning an exalted upper berth with mammals.

But don't laugh. Creationism may have its roots in indigenous American populism, but its exploiters and fundraisers are right-wing evangelicals who advance the literalism of Genesis as just one item in a comprehensive political program that would also ban abortion and return old-fashioned patriarchy under the guise of saving American families. Political programs demand political responses, but can we prevail without critical reason?

Index

adaptation, 35
 in evolution of human brain, 121–22
 by pandas, 23–24
 as result of natural selection, 59
adaptationism:
 evolutionary theory misstated in, 47–49
 fallacy in, 34–35
 Kevles', 43–45, 50
 in sociobiology, 30, 33, 40–41
Agnew, Spiro, 241
AIDS (acquired immune deficiency syndrome), 159
Aiken, George, 124
altruism, 30
amino acids, synthesis of, 235–36
analogous similarities, 221–22
animals:
 apes, 240
 behavior of, Bonner on, 65
 behavior of, Capra on, 224–25
 Bonner's hierarchy of, 65–66
 cultural forms in, 69
 female, Kelves on, 41–47
 language ability in, 242–43
 names of, 222
 pandas, 19–25
 proportion of insects among, 180
 in sociobiology, 33
anthropic principle, 48
apes, 240
Archaeopteryx, 233
Archilochus, 11
Archimedes, 107
Ardrey, Robert, 152
Aristotle, 162, 168
artificial intelligence, 215
astronomy, 212–13
Atlantic Ocean, 96
australopithecines, 109–10
Australopithecus, 152, 233

baboons, 42
Bacon, Francis, 217
Barnum, P. T., 240
Basin and Range, 97
Batocera, 180
Beagle (ship), 54, 124
beetles, 180
behavior:
 in adaptationism and sociobiology, 35
 animal, Capra on, 224–25
 cannibalism, 46–47
 human, gene-culture coevolution in, 108
 human flexibility in, 63–64
 methodological problems in study of, 119–20
 "nature-nurture" debate on, 112–14
 of pandas, 21–22
 sexual, Kevles on, 45
 sociobiology on, 28, 32–33, 67–68, 116–17, 120–21
 of solitaires (birds), 188
bias:
 in IQ scores, 127–30

bias *(continued)*
  Myrdal on, 216
  in science, 149–51, 230
big bang theory, 212–13
biogenetic law, 236
biological determinism, 147–49, 151–53
  gene-culture coevolution and, 107–12
  "nature-nurture" debate and, 112–14
biology:
  ecology and, 180–85
  Hutchinson's contributions to, 186
  interaction between culture and, 151–52
  in interactionism, 153
  E. E. Just's research in, 172–74
  "nature-nurture" debate in, 112–14, 148
  in sociobiology's hierarchy of sciences, 118
  at Wood's Hole Laboratory, 172
biotechnology, 229–32, 238
birds:
  cuckoos and redstart warblers, 45–46
  cultural transfer of songs by, 69
  evolution of, 233
  Galapagos finches, 54
  kiwis, 47
  pigeons, 26
  ring-billed gulls, 47
  solitaires, 187–88
blacks:
  assumptions of inferiority of, IQ scores of, 128–32
  Jensen on, 125–28
  E. E. Just, 169–79
Blyth, Edward, 54–60
body size, human, IQ scores and, 142
Bonner, John Tyler, 63–69
book reviews, 9–11
Boring, E. G., 132
brain, human:
  color perception and, 113–14
  computers *vs.*, 208–10, 214–15
  Darwin and A. R. Wallace on, 121–22
  evolution of, 48–49, 109–12
  flexibility in behavior and, 63
  gene-culture coevolution in, 108–9
  "handedness" of, 219
  Jensen on, 142
  sex differences in, 38, 144
brains, animal, 65
Brecht, Bertolt, 220
Brenner, Sidney, 220
Broca, Paul, 144
Browne, Sir Thomas, 183
Buffon, Georges, 51
Burt, Sir Cyril, 125, 134, 136, 147, 149
Butler, Samuel, 51–52
Butterfield, Herbert, 150
Button, Jemmy, 124

Cain, A. J., 40
Campanella, Roy, 179
cannibalism, 46–47
Capra, Fritjof, 216–28
Carboniferous Period, 80
Carter, Robert, 128
Cartesianism, 217–20, 226
Carver, George Washington, 173
Cascade Mountains, 93–94, 97
caste system, 116
Chambers, Robert, 52*n*
childrearing, 47, 71
China, pandas in, 19–25
chromosomes:
  cytogenetics of, 162
  DNA structure in, 157–58
  during meiosis, 163
  transposable elements in, 159
Coal Measures, 80–83
Cohen, Morris, 174
color perception, 113–14
communities (ecological), succession theory of, 182
computers, 208–10, 214–15
Comte, Auguste, 56
conflict:
  among animals, 224–25
  Capra on, 226
Cook, Robert E., 182*n*
Coon, Carleton, 152
Copernicus, Nicolaus, 214
correlation coefficients, 133
creationism, 245–46
Creighton, Harriet, 162

Crick, Francis, 122–23, 157–58
cuckoos, 45–46
cultural determinism, 152
cultural evolution, 33–34, 64–65, 70
  gene-culture coevolution in, 107–12
  sociobiology on, 39–40
culture:
  Bonner on, 63–65, 68–69
  genetic universals and, 112
  human differences among, 115–17
  in interactionism, 153
  in "nature-nurture" debate, 148
  shaped by individuals, 153
  in sociobiology, 118–20
"culturgens," 120
Cuvier, Georges, 82, 101–2
cytogenetics, 162

Darrow, Clarence, 124
Darsee, 190
Darwin, Charles, 161, 218
  "autobiography" of, 199–200
  Blyth and, 54–60
  Butler and, 51–52
  current evolutionary theory and, 232
  FitzRoy and, 124
  Freud on, 214
  on hierarchy of animals, 66
  on human cultural evolution, 110–11
  on human skull, 49
  on intermediate stages, 235
  Kelvin's criticisms of, 143–44
  T. H. Morgan on, 236–37
  Neo-Darwinism and, 67
  on pigeons, 26
  on progress, 211
  on sexual selection, 42
  Smith's economics in theories of, 103
  A. R. Wallace and, 121–22
Darwin, Erasmus, 51*n*
Darwin, Francis, 52
Darwinism:
  adaptationism in, 40, 44
  "cardboard," 50
  genetic modification in, 69–70
  in population ecology, 183
  "progress" not present in, 211
  randomness in, 213*n*
  Rifkin on, 231–36
  sexual variation in, 44–45
  of sociobiology, 27, 30, 32
  see also evolution
Dawkins, Richard, 66
Deffeyes, Ken, 97
De la Beche, Henry, 80–84, 86–87, 91–92
Descartes, René, 153
  Cartesianism and, 217–20, 226
Devonian controversy, 76, 78, 80–92
Devonshire (England), 80–81, 83
dialectical approach, 153–54
discrimination:
  by sex, 161, 163
  see also racism
diseases, 193–94
DNA (deoxyribonucleic acid), 157–59
Dobzhansky, Theodosius, 60
Doyle, Arthur Conan, 242
Dyson, Freeman, 201–7

earth:
  as organism, 102
  plate tectonics of, 94–99
eating, by pandas, 22–23
ecology, 180–86
  Capra on, 217, 227
education:
  intelligence testing in, 136
  reproduction rates and, 146–47
Eisley, Loren, 53–59
Eldredge, Niles, 37–38, 233–34
embryology:
  biogenetic law in, 236
  E. E. Just's contributions to, 172–73
  reductionism in, 176
Engels, Friedrich, 111–12
environment:
  in cultural determinism, 152
  ecological niches in, 183–85
  intelligence and, popular notions of, 145, 147
  IQ scores and, 130–32, 141–42
  shaped by organisms, 153

ESP (extrasensory perception), 241, 242, 245
eugenics, 146–47, 174
evolution:
  creationism and, 245–46
  during Devonian Period, 85
  of human brain, 121–22
  of human mind, 109–12
  Jastrow on, 210–12
  Kelvin's criticisms of, 143–44
  misstated in adaptationism, 47–48
  no hierarchical ladder in, 138–39
  of pandas, 23–24
  in population ecology, 183
  priority disputes in, 51–53, 60–61
  "progress" in, 204
  punctuated equilibrium in, 38
  Rifkin on, 231–36
  variation needed for, 167
  see also cultural evolution; Darwinism

factor analysis, 132–36
  Thurstone on, 136–40
fatherhood, 71
Fausto-Sterling, Anne, 37–39
females, human:
  education and reproduction by, 146–47
  roles of, 72
  sex discrimination against, 161, 163
  size of brains of, 144
  sociobiology on, 36
  E. O. Wilson on, 29
  Peter J. Wilson on, 71
  see also sex differences
females, nonhuman:
  Kevles on, 41–47
  Peter J. Wilson on, 71
Feuerbach, Ludwig, 154
final causes, 100
first principle component, 134–35
FitzRoy, Robert, 124
Flexner, Abraham, 175, 176
"flood geology," 245–46
Ford, Henry, 187
Fossey, Dian, 46
fossils:
  in Devonian controversy, 80–84, 86, 91
  in Old Red Sandstone (rock formation), 89–90
Freud, Sigmund, 214, 215

g (general intelligence), 133–42
Galapagos finches, 54
Galapagos Islands, 237
Gamow, George, 199
Gardner, Martin, 241–45
Gause, G. F., 182
Geller, Uri, 242, 244
gender differences, *see* sex differences
gene-culture coevolution, 107–12, 116
genes:
  different functions of, 159
  discovery of DNA structure of, 157–58
  natural selection on level of, 66–67
  transposed, McClintock's research on, 160–63
genetic engineering, 229–30, 238
genetics:
  of cultural evolution, sociobiology on, 39–40
  in Darwinism, 69–70
  discovery of DNA structure for, 157–58
  gene-culture coevolution and, 107–12
  of human behavior, sociobiology on, 67–68
  in interactionism, 153
  Jensen on, 125–28
  in politics of Singapore, 145–47
  in sociobiology, 118, 120–21
  transposable elements in, 159–63
  universals in, 112–15
genome, 158–59
Geological Society of London, 79, 86
geology:
  Devonian controversy in, 80–92
  Hutton in history of, 78–79, 99–103
  Noah's flood and, 245–46
  plate tectonics in, 95–98
  time scale used in, 76–78
giant pandas, 19–25

Gilbert, Sir William Schwenck, 75, 218
God:
  Blyth on, 60
  Hutton on, 100–101
  Jastrow on, 212–13
Goliath beetle, 180
gorillas, 46
granite, 82
Grassé, Pierre-Paul, 234
greywacke (rock), 80, 84
Gruber, H. E., 56, 57

Haeckel, Ernst, 110, 111
Haldane, J. B. S., 180
hedgehogs, 12
Herblock (Herbert Block), 9
heritability, 125–26
  popular notions of, 145, 147
Hobbes, Thomas, 237
Hobbs, William, 102–3
holism, 220–28
homologous similarities, 221
homosexuality, in ring-billed gulls, 47
Horace, 47
Horn, Henry, 185, 186
horses, evolution of, 235
Hrdy, Sarah B., 43
Hu Jinchu, 19*n*, 20
human nature, sociobiology on, 28, 115–17
humans:
  in anthropic principle, 48
  cultural differences among, 115–17
  culture of, 68–70
  diseases of, 193–94
  evolution of culture of, 64–65
  evolution of mind of, 109–12
  flexibility in behavior of, 63–64
  gene-culture coevolution in, 107–9
  Jastrow on, 210
  languages of, Dyson on, 205
  right-handedness in, 219
  skulls of, 49
  sociobiology of, 28–36, 67–68
Hutchins, Carleen, 185
Hutchinson, G. Evelyn, 180–88
Hutton, James, 78–79, 82, 98–103, 213
Huxley, Julian, 60
Huxley, Thomas Henry, 22, 52*n*, 149

incest, 112
indexes, 244
infanticide, 46
insects:
  beetles, 180
  mason wasp *(Monobia quadridens)*, 62
intelligence:
  artificial, 215
  brain size and, 144
  Jensen on, 125–28
  measurement of, 132–42
  popular assumptions about, 145
interactionism, 152–53
intermediate stages debate, 233, 235
IQ scores:
  environment and, 142–43
  factor analysis of, 132–36
  forms of bias in, 128–30
  heritability of, Jensen on, 125–28
  Jensen on, 130–32
  popular notions of heritability of, 147

Jastrow, Robert, 208–15
Jensen, Arthur, 124–33, 138–44, 147
"jumping genes," 160–63
Just, Ernest Everett, 169–79

Kamin, Leon, 147–49, 151–53
Keller, Evelyn Fox, 161, 164–68
Kelvin, Lord William Thomson, 143–44
Kerkut, Gerald, 234
Kevles, Bettyann, 41–47, 50
Kingsley, Charles, 149
kinship, Peter J. Wilson on, 70–71
Kitcher, Philip, 28–34, 39–40, 43
kiwis, 47
Koestler, Arthur, 226–27
Kohn, E. D., 55
Krause, Ernest, 51*n*
Kuhn, Thomas, 27

Lamarck, Jean Baptiste, 26, 51, 52*n*, 138
Lamarckianism, 70

Lamb, Charles, 10
language:
  animals' abilities for, 240, 242–43
  Dyson on, 205
Leakey, L. S. B., 200
Lee Kuan Yew, 145–47
Leguat, 188
Lewontin, Richard, 148, 151–53
Lillie, Frank R., 171, 172, 175–78
Limoges, C., 55
Lindeman, Raymond, 182–83
Loeb, Jacques, 175–76
Lotka, A. J., 182
Lull, Ramon, 243–44
Lumsden, Charles L., 28, 39–40, 107–23
Lyell, Sir Charles, 52*n*, 85, 102

McClintock, Barbara, 160–68
McPhee, John, 93–100
magicians, 242, 244
maize, 162, 167
males, human:
  differences between females and, 38–39
  sociobiology on, 36
males, nonhuman:
  childrearing by, 47
  sexual behavior of, 45
  sexual selection by, 41–42
Malthus, Thomas, 55, 56
Manning, Kenneth R., 169–71, 173–75, 178
Manouvrier, Léonce, 144
Marx, Groucho, 154
Marx, Karl, 154
mason wasp *(Monobia quadridens)*, 62
mating plugs, 45
Matthew, Patrick, 61
May, Robert, 185, 186
Mayr, Ernst, 55, 60
Mazlish, Bruce, 214, 215
measurement of intelligence, 132–42
Medawar, Peter, 85
medicine, L. Thomas on, 191–96
meiosis, 163
men, *see* males, human
Mendel, Gregor, 120, 121
Merton, Robert K., 52
methods:
  decline of written word and, 75–76
  dialectical approach to, 153–54
  in ecology, 181
  in geology, for measuring time, 81–82
  in IQ scores, 127–32
  in modern research laboratories, 190
  null results and, 37
  in population ecology, 183
  Rifkin on, 234
  of sociobiology, 117–120
  Peter J. Wilson's, 72
Micromalthus, 180
Mies van der Rohe, Ludwig, 32
Miller, Stanley, 235–36
mind, human:
  Dyson on, 206
  evolution of, 109–12
  in sociobiology, 118–19
  see also brain, human
Mivert, St. George Jackson, 233
Monobia quadridens (mason wasp), 62
Morgan, T. H., 236–37
Mt. Lassen, 94
Mt. Rainier, 93, 94
Mt. St. Helens, 93, 94
Murchison, Roderick Impey, 78, 80–87, 91–92
musical instruments, 185
Mussolini, Benito, 179
Myrdal, Gunnar, 216, 218

natural selection, 60
  in adaptationism and sociobiology, 34
  Blyth on, 57–58
  chance in, 213*n*
  Darwin's claim to, 55
  in evolution of human brain, 121–22
  helpful and harmful structures modified by, 49
  on level of genes, 66–67
  Matthew on, 61
  randomness and, 232
  Rifkin on, 233
  Rifkin on Morgan on, 236–37
  Smith's economics in, 103

Wallace and, 52
see also Darwinism; evolution
"nature-nurture" debate, 112–14, 148
Neo-Darwinism, 67
neoteny, 63
niches (ecological), 183–85
Noah's flood, 245–46
nuclear power, 204
nuclear weapons testing, 203
null results, 37

occultism, 241
Old Red Sandstone (rock formation), 89–90
Oppenheimer, J. Robert, 201, 203
organisms:
ecological niches for, 183–85
see also animals

paleontology, 210
Paleozoic Era, 79–80
pandas, 19–25
Pan Wenshi, 19*n*
paternoster lakes, 95
Pearl, Raymond, 174
Peckham, Robert F., 128
philanthropy, 174–76
physics, Capra on, 223–24
pigeons, Darwin on, 26
placebo effect, 193
plants, maize, 162, 167
plate tectonics, 94–99
Pope, Alexander, 207, 210
population ecology, 182, 183
Porter, Roy, 102
primary mental abilities (PMAs), 137–38
primates, Jastrow on, 212
principal components analysis, 134*n*–35*n*, 136–37
priority disputes, in evolution, 51–53, 60–61
production (systems) ecology, 182
punctuated equilibrium theory, 38, 233–34

Quetelet, Lambert, 56
quinarian system of taxonomy, 59

races (human):
IQ scores of, 128–32
Jensen on, 125–28
A. R. Wallace on, 121
racism, 179
E. E. Just's experiences with, 171–74, 177–78
Myrdal on, 216
randomness, in Darwinism, 213*n*, 232
Raup, David, 40
redstart warblers, 45–46
reductionism:
Capra's, 217, 224
Cartesian, 218–20
in discovery of DNA structure, 157
in embryology, 173, 176
in interactionism, 153
in sociobiology, 118
Reger, Max, 9–10
religion:
Brecht on, 220
Eastern, Capra on, 223
functions of, 122
E. O. Wilson and Lumsden on, 120
reproduction:
in classical Darwinism, 30
human, eugenic theories of, 146–47
E. E. Just's research on, 172–73
among pandas, 21
role of females in, 42
sexual, function of, 44–45
retroviruses, 159
reverse transcriptase, 159
Rifkin, Jeremy, 230–39
ring-billed gulls, 47
RNA (ribonucleic acid), 159
Robinson, Jackie, 179
Rockefeller Foundation, 174
rocks:
determining age of, 81–82
in Devonian controversy, 82–83
Hutton on, 103
in Old Red Sandstone (formation), 89–90
Paleozoic, 79–80
Rose, Steven, 148, 152–53
Rosenwald, Julius, 174
Rudwick, Martin J. S., 76, 78, 83, 84, 87–92
Russell, Bertrand, 199

Safire, William, 241
Saint-Hilaire, Geoffroy, 24–25
Sayers, Dorothy, 166
Schaller, George B., 19–25
Schopf, Tom, 40
Schweber, S. S., 56, 57
science:
  bias in and social functions of, 149–51
  creativity in, 103
  dialectical approach to, 153–54
  Freud on, 214
  McPhee and, 98
  philanthropy and, 174–76
  popular, 199–200
  racism in, 171–74, 177–78
  Rifkin on role of fact and theory in, 234
  search for God in, 213
  sex discrimination in, 161, 163
  social context of, 230
  sociobiology's hierarchy of, 118
sea urchins, 15
second principal component, 135
Sedgwick, Adam, 78
Sepkoski, Jack, 40
sex:
  discrimination based on, 161, 163
  Kevles on function of, 44–45
  selection by, Darwin on, 42
sex differences, 37
  in brain sizes, 144
  patterns in research on, 38–39
  sociobiology on, 29, 36
Shockley, William, 152
Silvers, Robert, 11
Simberloff, Dan, 40
simple structure axes, 137–40
Simpson, George Gaylord, 60, 235
Singapore, 145–47
skulls, human, 49
Smith, Adam, 56, 103
Smith, William, 78, 82
sociobiology, 27–28
  adaptationism in, 40–41, 48–50
  assumptions of, 117–23
  on cultural differences, 115
  gene-culture coevolution in, 107–12
  on human behavior, 67–68
  of humans, 28–36
  Kevles', 43
  on sex differences, 39
  E. O. Wilson and Lumsden on future of, 122–23
socioeconomic determinism, 230–31
soil, 101
solitaires (birds), 187–88
Sonneborn, Tracy, 163
space colonization, 203
spatial perception in women, 39
Spearman, Charles, 133–36
  Thurstone on, 136–40
species:
  of beetles, 180
  Blyth on, 57
  created by genetic engineering, 229–30
  Darwin on, 55
  diversity of, 181
  ecological niches for, 184–85
  in population ecology, 183
  punctuated equilibrium theory on, 233–34
  Rifkin on, 231–32
  succession theory of, 182
Stanford-Binet intelligence test, 131, 143
succession theory, 182
Sullivan, Sir Arthur Seymour, 75, 218
superposition, 81
systems (production) ecology, 182

taxonomy, quinarian system of, 59
Teilhard de Chardin, Pierre, 210, 221
Teller, Edward, 203
theory:
  current evolutionary theory, 232
  in ecology, 183–86
  in geology, 79
  Rifkin on, 234
  in sociobiology, 31
  succession theory, 182
Thomas, Lewis, 189–96
Thomas, Lewis, Sr., 189, 191, 194
Thurstone, L. L., 133, 136–40
time, geological, 76–78
  Hutton and, 99–100

McPhee on, 98
measurement of, 81–82
trichopterygids, 180

uniformitarianism, 85, 102

variation, 59, 167
vitalism, 173
volcanos, 93–94
Voltaire, 205–6
Volterra, V., 182
Vonnegut, Kurt, 229

Wallace, Alfred Russel, 52, 121–22
Washington, Booker T., 173
wasps, 62
Watson, James D., 157, 158
Werner, Abraham Gottlob, 82
Wernerian theory, 82
Whitman, Walt, 243
Williams, George, 49
Wilson, Edward O., 27–29, 35–36, 107–23
Kitcher on, 34, 39–40
Wilson, Peter J., 63, 64, 68–72
Wolong Natural Reserve (China), 20
women, *see* females, human
Woods Hole, Marine Biological Laboratory at, 169, 171–72, 176–77
World War II, 202–3
Wright, Sewall, 66

yin and yang, 223
Young, J. Z., 11

Zhu Jing, 19*n*
Zirkle, C., 58
zoos, 25, 46
Zuckerman, Solly, 42